ENERGY HARVESTING AUTONOMOUS SENSOR SYSTEMS

Design, Analysis, and Practical Implementation

T0225537

ENERGY HARVESTING AUTONOMOUS SENSOR SYSTEMS

Design, Analysis, and Practical Implementation

Yen Kheng Tan

CRC Press
Taylor & Francis Group
Boca Raton   London   NewYork

CRC Press is an imprint of the
Taylor & Francis Group, an **informa** business

CRC Press
Taylor & Francis Group
6000 Broken Sound Parkway NW, Suite 300
Boca Raton, FL 33487-2742

First issued in paperback 2017

© 2013 by Taylor & Francis Group, LLC
CRC Press is an imprint of Taylor & Francis Group, an Informa business

No claim to original U.S. Government works

Version Date: 20121214

ISBN 13: 978-1-4398-9273-2 (hbk)
ISBN 13: 978-1-138-07409-5 (pbk)

Visit the Taylor & Francis Web site at
http://www.taylorandfrancis.com

and the CRC Press Web site at
http://www.crcpress.com

Contents

Preface

What This Book Is About

With the recent advances in wireless communication technologies, sensors and actuators and highly integrated microelectronics technologies, wireless sensor networks (WSNs) have gained worldwide attention by facilitating the monitoring and control of physical environments from remote locations, which can be difficult or dangerous to reach. WSNs represent a significant improvement over wired sensor networks with the elimination of the hardwired communication cables and associated installation and maintenance costs. The possible uses of WSNs for real-time information in all aspects of engineering systems are virtually endless, from intelligent building control to health care systems, environmental control systems, and more. As electronic hardware circuitries become cheaper and smaller, more and more of these WSN applications are likely to emerge, particularly as these miniaturized wireless sensor nodes offer the opportunity for electronic systems to be embedded unobtrusively into everyday objects to attain a "deploy-and-forget" scenario.

In the great majority of autonomous sensor nodes in WSNs, electrical energy that is necessary for their operation is provided primarily by batteries. Batteries take up a significant fraction of the total size and weight of the overall system. Moreover, they are also the weakest link but yet the most expensive part of the system. Another important factor to be considered is the requirement for the proper maintenance of batteries, with the need either to replace or recharge them. This is a serious limitation of WSNs, in which there are dozens or hundreds of sensor nodes with batteries to maintain. Generally, the highest reported energy for present-day battery technologies ranges around 3.78 kJ/cm^3 [1], which implies that an ultralow-power miniaturized wireless sensor node with a volumetric size of 1 cm^3 operating at an average power consumption of 100 μW to have a 10-year life span needs a battery as large as 10 cm^3. Thus, energy supply is one of the major bottlenecks for the lifetime of the sensor node and is constrained by the size of the battery.

The major hindrances of the "deploy-and-forget" nature of WSNs are the their limited energy capacity and the unpredictable the lifetime performance of the battery. To overcome these problems, *energy harvesting (EH)/scavenging*, which harvests/scavenges energy from a variety of ambient energy sources and converts it into electrical energy to recharge the batteries, has emerged as a promising technology. With the significant advancement in microelectronics, the energy and therefore the power requirement for sensor nodes continues to decrease from a few milliwatts to a few tens of microwatts. This paves the way for a paradigm shift from the battery-operated conventional WSN, which solely relies on batteries, towards a truly self-autonomous and sustainable energy harvesting wireless sensor network (EH-WSN). Various types of EH systems and their respective main components (i.e., energy harvester [source], power management circuit, energy storage device, and wireless sensor node [load]) are investigated and analyzed in this work. EH systems, based on wind energy harvesting (WEH), thermal energy harvesting (TEH), vibration energy harvesting (VEH), solar energy harvesting (SEH), hybrid energy harvesting (HEH), and magnetic EH, are designed to suit the target applications regarding ambient conditions and event/task requirements and then implement into hardware prototypes for proof of concept. To optimize these EH systems, several different types of power-electronic–based management circuits, such as an active alternating current-direct current (AC-DC) converter, DC-DC converter with maximum power point tracking (MPPT), energy storage and latching circuit, and more have been introduced.

Like any of the commonly available renewable energy sources, WEH has been widely researched for high-power (greater than a few megawatts level) applications. However, few research works on the small-scale WEH, which are used to power small autonomous sensors, can be found in the literature. A small-scale WEH system has the problems of low-magnitude generated output voltage and low harvested electrical power; as such, they pose severe constraints on the circuit design of the power management unit of the WEH wireless sensor node. To overcome the problems mentioned, an optimized WEH system that uses an ultralow-power management circuit with two distinct features is proposed: (1) an active rectifier using a metal-oxide semiconductor field-effect transistor (MOSFET) for rectifying the low-amplitude AC voltage generated by the wind turbine generator (WTG) under a low wind speed condition efficiently; and (2) a DC-DC boost converter with a resistor emulation algorithm to perform MPPT under varying wind speed conditions. As compared to the conventional diode-bridge rectifier, it is shown that the efficiency of the active rectifier has been increased from 40% to 70% due to the significant reduction in the on-state voltage drop (from 0.6 V to 0.15 V) across each pair of MOSFETs used. The proposed robust low-power microcontroller-based resistance emulator is implemented with a closed-loop resistance feedback control to ensure close impedance matching between the source and the load, resulting in efficient power conversion. From the experimental test results obtained, an average electrical power of 7.86 mW is harvested by the optimized WEH system at an average wind speed of 3.62 m/s, which is almost four times higher than the conventional EH method without using MPPT.

For space constraint applications where a small-scale WEH system needs to be as small as possible and highly portable, this type of conventional, large, and bulky WTG is not that suitable. As such, a novel method to harvest wind energy using piezoelectric material lead zirconate titanate (PZT) has been presented. The overall size of the proposed PZT structure is much smaller compared to the WTG. Energy harvested from the piezoelectric-based wind energy harvester is first accumulated and stored in a capacitor until there is sufficient stored energy to power the sensor node; a trigger signal is then initiated to release the stored energy in the capacitor to the wind speed sensor node. Experimental results show that the harvested stored energy of 917 μJ is used to detect wind speed beyond a certain threshold level of 6.7 m/s for an early warning storm detection system.

In some places where a wind energy source is not available, TEH from ambient heat sources with low temperature differences have recently received great attention but has been impeded by the challenges of low energy conversion efficiency, inconsistency, low output power due to temperature fluctuation, and higher cost. To supplement the TEH scheme, an efficient power management circuit that could maximize power transfer from the thermal energy source to its connected electronic load is desirable over a wide range of operating conditions. In this work, a DC-DC buck converter with a resistor emulation-based maximum power point (MPP) tracker is presented for an optimal TEH scheme in sustaining the operation of wireless sensor nodes. From the experimental test results, an average electrical power of 629 μW is harvested by the optimized TEH system at an average temperature difference of 20 K, which is almost two times higher than the conventional EH method without using an MPPT scheme.

Electrical cables that are used in residential and industrial buildings to connect an appliance to a control switch on the wall have been a cause of nuisance as well as led to higher installation costs. Undesirable recabling implications may also arise should the cable become faulty over time. To overcome the problem, a batteryless and wireless remote controller is proposed to switch electrical appliances such as lights and fans on/off in a wireless manner. In this work, two types of piezoelectric-based VEH systems are presented to harvest impact or impulse forces from a human pressing a button or using a switch. By depressing (1) the piezoelectric push-button igniter or (2) the prestressed piezoelectric diaphragm material, electrical energy is generated and stored in the capacitor. Once sufficient energy is harvested, the batteryless and wireless remote controller is powered up for operation.

An EH system itself has an inherent problem: the intermittent nature of the ambient energy source. The operational reliability of the wireless sensor node may be compromised due to unavailability of the ambient energy source for a prolonged period of time. To augment the reliability of the wireless sensor node operation, two types of HEH approaches are investigated. A hybrid WEH and SEH scheme is proposed to harvest simultaneously from both energy sources to extend the lifetime of the wireless sensor node. When the two energy sources with different characteristics are combined, there is bound to be impedance mismatch between the two different sources and the load. Hence, each energy source has its own power management unit to maintain at its respective MPP. The WEH subsystem uses the resistor emulation technique, while the SEH subsystem uses the constant voltage technique for MPP operation. Experimental results show that an average electrical power of 22.5 mW is harvested by the optimized HEH system at an average wind speed of 4 m/s and an average light irradiance of 80 W/m^2, which is almost three times higher than the single wind-based energy source.

In another HEH research work, a hybrid of indoor ambient light and a TEH scheme that uses only one power management circuit to condition the combined output power harvested from both energy sources is proposed to extend the lifetime of the wireless sensor node. By avoiding the use of individual power management circuits for multiple energy sources, the number of components used in the HEH system are reduced, and the system form factor, cost, and power losses are thus reduced. An efficient microcontroller-based ultralow-power management circuit with fixed voltage reference-based MPPT is implemented with a closed-loop voltage feedback control to ensure near maximum power transfer from the two energy sources to its connected electronic load over a wide range of operating conditions. From the experimental test results, an average electrical power of 621 μW is harvested by the optimized HEH system at an average indoor solar irradiance of 1010 lux and a thermal gradient of 10 K, which is almost triple that obtained with a conventional single thermal-based energy source.

Other than EH, this work also demonstrates an alternative means to remotely power low-power electronic devices through a wireless power transfer (WPT) mechanism. The WPT mechanism uses the concept of inductive coupling (i.e., harvesting the stray magnetic energy in power lines to transfer electrical power without any physical connection). The AC voltage and current in the power lines are 230 V and 1 to 4 A, respectively. Experimental results show that the implemented magnetic energy harvester is able to harvest 685 μJ of electrical energy from the power lines to energize the radio-frequency (RF) transmitter to transmit 10 packets of 12-bit encoded digital data to the remote base station in a wireless manner. To extend the WPT distance, self-resonating coils, operating in a strongly coupled mode, are demonstrated. Experimental results show that the WPT system is capable of delivering wireless output power up to 1 W at an efficiency of 51% over a separation distance of 20 cm to power a small lightbulb.

Until this stage, the proof of concepts for the developed EH prototypes have been demonstrated. The performance of the EH systems in powering wireless sensor nodes are investigated and tested under various operating conditions simulated in the laboratory. In addition, the EH prototypes are optimized according to their designed applications. However, in reality, the environmental conditions of the deployment area are not as ideal as in the laboratory environment. Therefore, the next stage of this EH research, which is considered future work, is to carry out a series of application-specific field trials to evaluate the performance of EH systems under real-life deployment conditions for a prolonged period of time. For the EH mechanism to be successful, an overall system optimization with respect to energy consumption, taking into account the duty-cycling operation of the WSNs for the entire chain (i.e., from sensing the environmental parameter to transmitting and delivering the sensed parameter reliably), is to be investigated. This part of the work is beyond the scope of this book and therefore is proposed as future research work.

1

Introduction

The rapid growth in demand for computing everywhere has made the computer a pivotal component of human's daily lives [2]. Whether we use the computers to gather information from the Web, for entertainment, or for running a business, computers are noticeably becoming more widespread, mobile, and smaller in size. What we often overlook is the presence of those billions of small pervasive computing devices around us that provide intelligence integrated into the real world *smart environments* [2] to help us solve some crucial problems in the activities of our daily lives. While we were asleep at the switch learning to "plus one" on Google, the Internet of Things (IOT) just exceeded the number of people that reside on the planet. Beyond just smartphones and tablets, the number of "things" that connect to the Internet will only continue to scale as the growing number of connected gizmos and appliances—and even cows—are coded and catalogued to send messages to the Web. Dave Evans at Cisco noted that "there are more devices tapping into the Internet than people on Earth to use them." How is that possible? Well, an infographic the firm just published, as shown in Figure 1.1, provides us insight with a visual snapshot of the increase in things connected to the Internet—and how they will serve us in the very foreseeable future.

By the year 2020, as can be seen in Figure 1.1, there will be 50 billion of these things around us. To achieve this vision of the smart environment with *pervasive computing*, also known as *ubiquitous computing*, many such miniaturized computing devices will be integrated in everyday objects and activities to enable better human-computer interaction. These computational devices, which are generally equipped with sensing, processing, and communicating abilities, are known as *wireless sensor nodes*. When these wireless sensor nodes are connected, they form a network called the *wireless sensor network (WSN)*, as illustrated in Figure 1.2.

1.1 Motivation of Wireless Sensor Networks (WSNs)

The postmodern era is a world where everything including people is connected like the illustration given in Figure 1.3. With surrounding close to invisibly small smart computing devices and sensors embedded in everyday

FIGURE 1.1
Internet of Things exceeds the Internet of People.

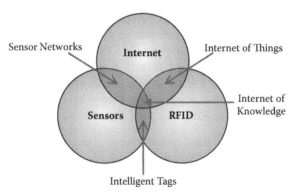

FIGURE 1.2
Overlap between WSN and IOT. (RFID: radio-frequency identification.)

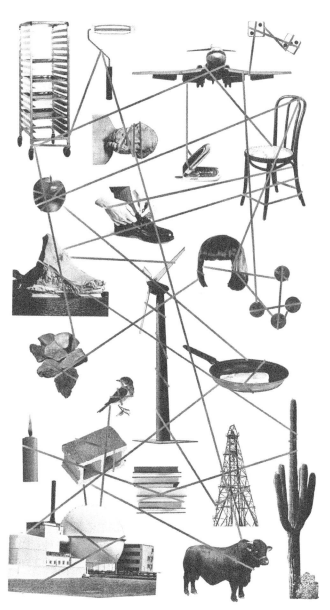

FIGURE 1.3
When everything connects.

ambient objects, environments are able to recognize and respond to the presence and behaviour of any individual in a personalized and relevant way. With the recent advances in wireless communication technologies, sensors and actuators, and highly integrated microelectronics technologies, WSNs have gained worldwide attention to facilitate monitoring and controlling of physical environments from remote locations that could be difficult or dangerous to reach. In the Massachusetts Institute of Technology (MIT) *Technology Review* magazine of innovation published in February 2003 [3], the editors identified WSNs as the first of the top 10 emerging technologies that will change the world.

Across many industries, products and practices are being transformed by these networked communicating sensors and computing intelligence. The smart industrial gear includes jet engines, bridges, and oil rigs that alert their human minders when they need repairs, before equipment failures occur. Computers track sensor data on operating performance of a jet engine or slight structural changes in an oil rig, looking for telltale patterns that signal coming trouble. Sensors on fruit and vegetable cartons can track location and sniff the produce, warning in advance of spoilage so shipments can be rerouted or rescheduled. Computers pull GPS data from railway locomotives, taking into account the weight and length of trains, the terrain, and turns to reduce unnecessary braking and curb fuel consumption by up to 10%.

1.1.1 Architecture of WSNs

WSNs represent a significant improvement over wired sensor networks with the elimination of the hardwired communication cables and associated installation and maintenance costs. An overview of these network systems is illustrated in Figure 1.4. The architecture of a WSN typically consists of

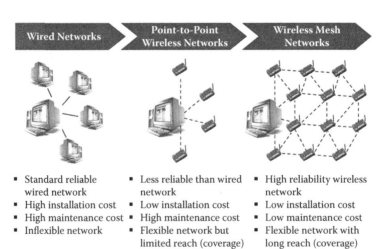

FIGURE 1.4
Comparison of WSN and IOT.

multiple pervasive sensor nodes, sink, public networks, manager nodes, and end user [4]. Many tiny, smart, and inexpensive sensor nodes are scattered in the targeted sensor field to collect data and route the useful information back to the end user. These sensor nodes cooperate with each other via a wireless connection to form a network and collect, disseminate, and analyze data coming from the environment. To ensure full connectivity, fault tolerance, and a long operational life, WSNs are deployed in an ad hoc manner, and the networks use multihop networking protocols to obtain real-world information and perform control ubiquitously [5]. As illustrated in Figure 1.5, the data collected by *node A* is routed within the sensor field by other nodes. When the data reaches the boundary, *node E*, it is then transferred to the sink. The sink serves as a gateway with a higher processing capacity to communicate with the task manager node. The connection between the sink and task manager node is the public network in the form of the Internet or a satellite. Once the end user receives the data from the task manager node, some processing actions are then performed on the received data.

In Figure 1.5, the sink is essentially a coordinator between the deployed sensor nodes and the end user, and it can be treated like a gateway node. The need of a sink in WSN architecture is due to the limited power and computing capacity of each of the wireless sensor nodes. The gateway node, typically powered by the readily available power source from the AC (alternating current) main, is equipped with a better processor and sufficient memory space that it is able to provide the need for extra information processing before data is transferred to the final destination. The gateway node can therefore share the loadings posed on the wireless sensor nodes and hence prolong their working lifetime. To understand how data is communicated within the sensor nodes in a WSN as shown in Figure 1.5, the protocol stack model of the WSN as shown in Figure 1.6 is investigated. With this understanding, the energy-hungry portions of the wireless sensor node can be identified, and

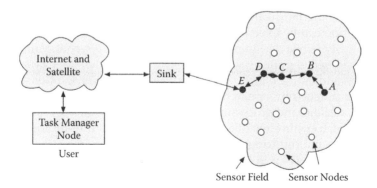

FIGURE 1.5
Architecture of a WSN to facilitate smart environments. (From I.F. Akyildiz, W.L. Su, S. Yogesh, and C. Erdal, "A survey on sensor networks," *IEEE Communications Magazine*, vol. 40, no. 8, pp. 102–114, 2002 [4].)

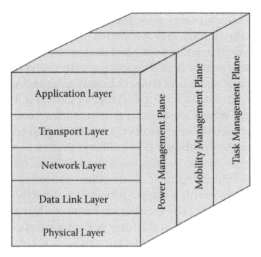

FIGURE 1.6
Sensor networks protocol stack. (From I.F. Akyildiz, W.L. Su, S. Yogesh, and C. Erdal, "A survey on sensor networks," *IEEE Communications Magazine*, vol. 40, no. 8, pp. 102–114, 2002 [4].)

then the WSN can be redesigned accordingly for lower power consumption. To start the basic communication process, consists of sending data from the source to the destination. Primarily, it is the case of two wireless sensor nodes wanting to communicate with each other. The sensor node at source generates information, which is encoded and transmitted to the destination, and the destination sensor node decodes the information for the user. This entire process is logically partitioned into a definite sequence of events or actions, and individual entities then form layers of a communication stack. The WSN protocol stack [4] shown in Figure 1.6 consists of five network layers: physical (PHY) (lowest), data link, network, transport, and application (highest) layers.

Starting from the lowest level, the PHY layer receives and transfers data collected from the hardware. It is well known that long-distance wireless communication can be expensive in terms of both energy and implementation complexity. While designing the PHY layer for WSNs, energy minimization is considered significantly more important than the other factors, like propagation and fading effects. Energy-efficient PHY layer solutions are currently being pursued by researchers to design for tiny, low-power, low-cost transceiver, sensing, and processing units [6]. The next-higher layer is the data link layer, which ensures reliable point-to-point and point-to-multipoint connections for the multiplexing of data streams, data frame detection, medium access, and error control in the WSN. The data link layer should be power aware and at the same time minimize the collisions between neighbours' signals because the environment is noisy and sensor nodes themselves are highly mobile. This is also one of the layers in the WSN whereby power saving modes of operation can be implemented. The most obvious means of power conservation is to turn the transceiver off when it is not required. By using a random wake-up schedule during the connection phase and by turning the radio off during idle time slots, power conservation can be achieved. A dynamic power management scheme for WSNs has been discussed [7]; five power-saving modes were proposed, and intermode transition policies were investigated.

The network layer takes care of routing the data supplied by the transport layer. In WSN deployment, the routing protocols in the network layer are important because an efficient routing protocol can help to serve various applications and save energy. By setting appropriate energy and time delay thresholds for data relay, the protocol can help prolong the lifetime of sensor nodes. Hence, the network layer is another layer in the WSN to reduce power consumption. The transport layer helps to maintain the flow of data if the sensor network application requires it. Depending on the sensing tasks, different types of application software can be built and used on the application layer. In contrast to traditional networks that focus mainly on how to achieve high quality-of-service (QoS) provisions, WSN protocols tend to focus primarily on power conservation and power management for sensor nodes [7, 8] as well as the design of energy-aware protocols and algorithms for WSNs [5, 9] to reduce the power consumption of the overall wireless sensor network. By doing so, the lifetime of the WSN can be extended.

However, there must be some embedded trade-off mechanisms that give the end user the option of prolonging the WSN lifetime but at the cost of lower throughput or higher transmission delay. Conversely, the power consumption of the WSN can be reduced by sacrificing the QoS provisions, that is, by lowering the data throughput or having a higher transmission delay. Among the several challenging requirements posed on the design of the underlying algorithms and protocols of the WSNs, it is well known among academia as well as industry [10–12] that energy constraint is one of the most significant challenges in the WSN research field [13]. The functionalities of the WSN are highly dependent on the amount of energy that is available to be expended by each sensor node in the network. As such, the energy constraint challenge of WSN is substantial enough to be investigated and discussed in this book. It is a multiobjective optimization problem concerning various WSN parameters like QoS, transmission delays, lifetime, energy, and more.

In 2000, two standards groups, ZigBee, a HomeRF spinoff, and IEEE (Institute for Electrical and Electronics Engineers) 802 Working Group 15, combined efforts to address the need for low-power, low-cost wireless networking in the residential and industrial environments. Furthermore, the IEEE New Standards Committee (NesCom) sanctioned a new task group to begin the development of a Low Rate-Wireless Personal Area Network (LR-WPAN) standard, to be called 802.15.4. The goal of this group is to provide

TABLE 1.1

Basic Parameters of IEEE 802.15.4

Property	Range
Raw data rate	868 MHz: 20 kb/s; 915 MHz: 40 kb/s; 2.4 GHz: 250 kb/s
Range	10–20 m
Channel access	CSMA-CA and slotted CSMA-CA
Channels	868/915 MHz: 11 channels; 2.4 GHz: 16 channels
Frequency band	Two PHYs: 868 MHz/915 MHz and 2.4 GHz
Addressing	Short 8 bit or 64 bit IEEE
Latency	Down to 15 ms

a standard with ultralow complexity, cost, and power for low-data-rate wireless connectivity among inexpensive fixed, portable, and moving devices. The scope of Task Group 4 is to define the PHY and media access control (MAC) layer specifications. IEEE 802.15.4 has some basic devices. These devices can be a reduced-function device (RFD) or a full-function device (FFD). The RFD can talk only to an FFD, but the FFD can operate in three modes, serving as a personal-area network (PAN) coordinator, a coordinator, or a device. An FFD can talk to an RFD or other FFDs. An RFD can be used in extremely simple applications, such as a light switch or a passive infrared sensor. Most of the time, RFDs will not have many data to send; hence, they communicate with the FFD occasionally. They need fewer resources and minimum energy. Two or more devices within a personal operating space (POS) communicating with all the same physical channels constitute a Wireless Personal Area Network (WPAN) . However, a network will have at least one FFD operating as the PAN coordinator. The basic parameters of IEEE 802.15.4 are presented in Table 1.1.

The layout of the IEEE 802.15.4 blocks is defined based on the Open System Interconnection (OSI) seven-layer model. Separation of the LR-WPAN into blocks helps to understand the protocol. These blocks are called layers. The higher layers receive their information from the following lower-level layers. The interfaces between the layers serve to define the logical links. An LR-WPAN device consists of PHY and MAC sublayers. A PHY sublayer has the radio-frequency (RF) transceiver along with its low-level control mechanism, an MAC sublayer provides access to the physical channel for all types of transfer. The overview of the IEEE 802.15.4 architecture is depicted in Figure 1.7.

The upper layers consist of a network layer and the application layer. An application layer provides the intended function of the device, and the network layer provides network configuration and message routing. An IEEE 802.2TM Type 1 logical link control (LLC) can access the MAC sublayer through the service-specific convergence sublayer (SSCS).

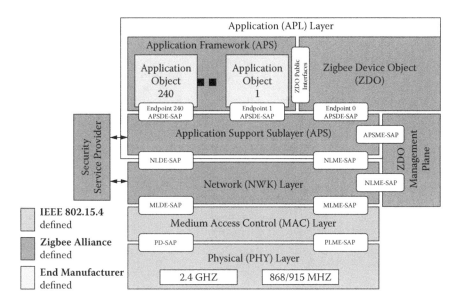

FIGURE 1.7
Architecture of IEEE Standard 802.15.4.

1.1.2 Applications of WSNs

The WSNs can be used in virtually any environment, even where wired connections are not possible, the terrain is inhospitable, or physical placement of the sensors is difficult. Besides that, WSNs enable autonomous monitoring of physical quantities over large areas on a scale that would be prohibitively expensive to accomplish with human beings. These attractive features promote the potential of WSNs for more application areas. There have been many applications suggested for WSNs in the literature, and they can be roughly classified into three categories as suggested by Culler, Estrin, and Srivastava [14]:

* Monitoring space
* Monitoring entities
* Monitoring the interactions of entities with each other and the encompassing space

The first classification includes environmental monitoring, indoor climate control, and military and space surveillance. In military applications, for example, a large quantity of the pervasive computing devices could be deployed over a battlefield to detect enemy intrusion instead of manually deploying the land mines for battlefield surveillance and intrusion detection [10]. The second classification includes condition-based equipment maintenance, medical health diagnostics, vehicle safety, urban terrain mapping, and structural

* Medical and Health Monitoring

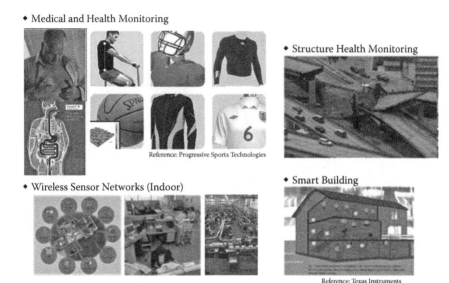

Reference: Progressive Sports Technologies

* Structure Health Monitoring

* Wireless Sensor Networks (Indoor)

* Smart Building

Reference: Texas Instruments

FIGURE 1.8
Examples of WSN application areas.

monitoring where pervasive computing devices are deployed to detect for any damage in buildings, bridges, ships, and aircraft [15]. The most dramatic applications fall under the third classification; these involve monitoring complex interactions, including wildlife habitats, disaster management, emergency response, asset tracking, and manufacturing process flow. A summary of these applications is shown in Figure 1.8.

More of these WSN applications are likely to emerge as electronic hardware circuitries become cheaper and smaller, and these miniaturized wireless sensor nodes offer the opportunity for electronic systems to be completely connected, intuitive, effortlessly portable, constantly available, and embedded unobtrusively and pervasively into everyday objects to attain a deploy-and-forget scenario.

1.1.3 Wireless Sensor Nodes of WSNs

The WSNs, based on the collaborative efforts of a large number of sensor nodes distributed throughout an area of interest, have been proven by many researchers as good candidates to provide economically viable solutions for a wide range of applications. These sensor nodes are coordinated based on some network topologies to cooperate with one another within the WSNs to satisfy the application requirements. Each sensor node monitors its local environment, locally processing and storing the collected data so that other

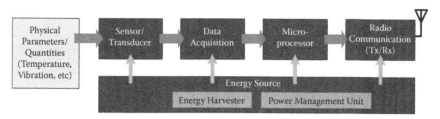

FIGURE 1.9
Block diagram of a wireless sensor node.

sensor nodes in the network can use it. As illustrated in Section 1.1.1, network nodes share this information via a wireless communication link. Because of the great potential of WSNs, many groups around the world have invested research efforts and time in the design of sensor nodes for their specific applications. These include Berkeley's Mica motes [16], PicoRadio projects [17], MIT's μAmps [18], as well as many others. In addition, the TinyOS project [19] provides a framework for designing flexible distributed applications for data collection and processing across the sensor network. All of these sensor nodes have similar goals, such as small physical size, low power consumption, and rich sensing capabilities. A block diagram of the wireless sensor node of a WSN is shown in Figure 1.9. The sensor node typically consists of four subunits: the sensor itself, data acquisition circuit, local microcontroller, and radio communication block. Some examples of a wireless sensor node are given in Figure 1.10.

Referring to Figure 1.9, the sensor/transducer converts an environmental parameter such as temperature, vibration, humidity, and the like to an electrical signal. A data acquisition circuit is incorporated in the sensor node to realize amplification and preprocessing of the output signals from sensors, for example, conversion from analog to digital form and filtering. The conditioned signals are then processed and stored in the embedded microcontroller for other sensor nodes in the network to use them. Other than data processing, the microcontroller also provides some level of intelligence such as time scheduling to the sensor node. To enable the sensor node to communicate with its neighbour node or the base station in a wireless manner, a radio communication block such as shown in Figure 1.9 is included. All four subunits of the sensor node are power sink modules, and they need to consume electrical energy from the power source in order to operate. Because of that, the wireless sensor node would consume all the energy stored in the battery after some time, and the sensor node would then go into an idle state. Once the percentage of nodes that have not terminated their residual energy falls below a specific threshold, which is set according to the type of application (it can be either 100% or less) [20], the operational lifetime of the WSN ends.

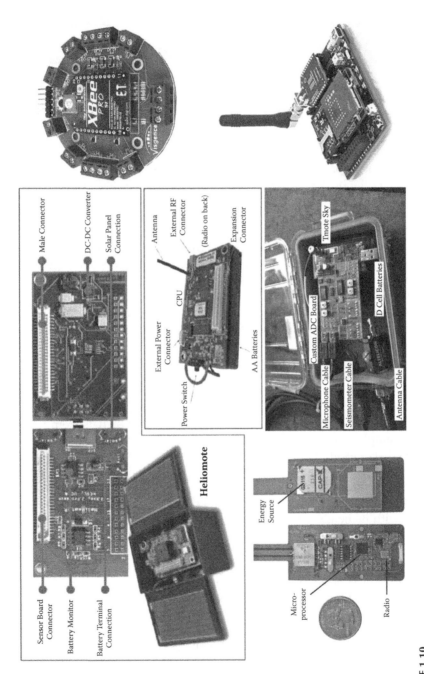

FIGURE 1.10
Examples of wireless sensor nodes.

1.2 Problems in Powering Wireless Sensor Nodes

As the WSN becomes dense with many sensor nodes, the problems in powering the wireless sensor nodes, namely (1) high power consumption of sensor nodes and (2) limitation of energy sources for sensor nodes becomes critical and is even worse when one considers the prohibitive cost of providing power through wired cables to them or replacing batteries. Furthermore, when the sensor nodes must be extremely small, as tiny as several cubic centimetres, to be conveniently placed and used, such small volumetric devices are very limited in the amount of energy that the batteries can store, and there would be severe limits imposed on the nodes' lifetime powered by the miniaturized battery, which is meant to last the entire life of the node.

1.2.1 High Power Consumption of Sensor Nodes

Based on the breakdown of a wireless sensor node illustrated in Section 1.1.3, the information about how much electrical power a sensor node consumes during operation is determined. The power consumed by each individual component (i.e., microcontroller, radio, logger memory, and sensor board in a sensor node) is tabulated in Table 1.2. It can be observed from Table 1.2 that all the components in the sensor node consume a milliwatt level of power during the active mode of operation and then drop to microwatts of power when in sleep or idle mode. If the sensor node is set to operate at full duty cycle, that is, 100%, the current and therefore power consumption of the sensor node would be as high as 30 mA.

For most practical WSN applications described in Section 1.1.2, *duty cycling* of the sensor node's operation is a common method discussed in the literature [12–14] to reduce its power consumption and therefore extend the lifetime of the WSN. During one operational cycle, the sensor node remains active for a brief period of time before going into the sleep mode. During the sleep period, the current consumption of the sensor node is typically in the microamper range as opposed to the active period in the milliamper range. This results in the sensor node drawing very little current, in the range of 2 to 8 μA for the majority of the time, and a short duration of current spikes in the range of 5 to 17 mA while sensing, processing, receiving, and transmitting data as illustrated in Table 1.2. For the Xbow sensor node given in Table 1.2, operating at a duty cycle of 1% as opposed to 100%, the average current consumption of a node with a supply voltage of 3 V is significantly reduced from 30 mA to around 0.3 mA. Referring to Figure 1.11, it is clearly illustrated that there is a drastic difference in the battery life of the wireless sensor node operating at duty cycles of 1% and 100%. This shows the need for duty cycling during the operation of the sensor node.

Even with the help of duty cycling in the operation of the sensor node, the power consumed by the wireless sensor node has been decreased by around 75 times, but the power density of its battery is still not high enough to support

TABLE 1.2

Battery Life Estimation for an Xbow Sensor Node Operating at 100% and 1% Duty Cycles

	System Specifications				
Currents			**Duty Cycles**		
	Value	Units	Model 1	Model 2	Units
Micro Processor (Atmega128L)					
Current full operation	8	mA	100	1	%
Current sleep	8	μA	0	99	%
Radio					
Current in receive	16	mA	75	0.75	%
Current xmit (3dB)	17	mA	25	0.25	%
Current sleep	1	μA	0	99	%
Logger					
Write	15	mA	0	0	%
Read	4	mA	0	0	%
Sleep	2	μA	100	100	%
Sensor Board					
Current (full operation)	5	mA	100	1	%
Current sleep	5	μA	0	99	%
Computed Average Current Consumed (mA)			**Model 1**	**Model 2**	
uP			8.0000	0.0879	
Radio			16.2500	0.1635	
Flash Memory			0.0020	0.0020	
Sensor Board			5.0000	0.0550	
Total current (ma) used			29.2520	0.3084	

Source: Crossbow Technology Inc., "MPR-MIB Users Manual," Crossbow Resources, Revision A, 2007 [21].

the high power consumption of the sensor node for a long period of time. Based on an AA alkaline battery of 3000 mA-hr, it is illustrated in Figure 1.11 that the lifetime of the battery powering the sensor node is computed to last at most 1.1 years. After which, without battery replacement, the sensor node can be considered an expired node. This is even worse for the case of a coin-type 250-mA-hr alkaline battery, which is smaller in size than the AA battery. Referring to Figure 1.11, the coin battery can only sustain the operation of the sensor node for at most 1 to 2 months. Clearly, a lifetime of a year or so or even less for the wireless sensor node to operate is far from sufficient because the duration of the node's operation could last for several years for the WSN to be useful in practical situations. This is a serious limitation to computing paradigms like ubiquitous computing or sensor networks, in which there are dozens or hundreds of small self-autonomous sensor node systems with batteries to maintain.

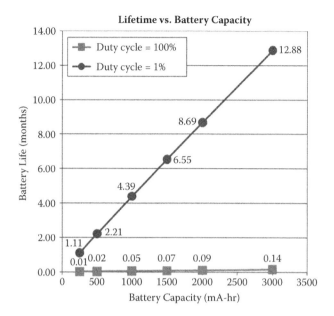

FIGURE 1.11
Expected battery life versus system current usage and duty cycle. (From Crossbow Technology Inc., "MPR-MIB Users Manual," Crossbow Resources, Revision A, 2007 [21].)

Another issue associated with the high power consumption of a wireless sensor node is the need for equivalently high power density energy storage or energy source. As can be seen in Figure 1.12, the instantaneous current drawn by the sensor node communicating with the gateway is as high as 0.35 A. This means that for a 3-V battery supply, 1 W of electrical power is required at the first instant to wake up the sensor node, as illustrated in Figure 1.12. Similarly, for the subsequent operations of the sensor node, that is, transmit (CSMA/CA TX) and receive (CSMA/CA RX), the sensor will demand 1 W of electrical power from the power source to complete its RF communication with its base station or gateway. Under this circumstance, the powering of the sensor node becomes a potential critical problem to be addressed.

1.2.2 Limitation of Energy Sources for Sensor Nodes

In many application scenarios, the lifetime of the sensor node typically ranges from 2 to 10 years depending on the requirement of the specific application. For the case of deploying sensor nodes on a mountain to detect the thickness of the ice on the mountain, it will take years for the melting process to be measurable. Hence, the lifetime of the sensor nodes must be several years before they go into an idle state. If that is the case, energy supply is one of the major bottlenecks to be addressed. Many different types of energy storage technologies are already available; as illustrated in Figure 1.13, alkaline/rechargeable batteries and supercapacitors are the most portable and popular energy supply option for powering the sensor nodes in WSNs. Batteries convert stored

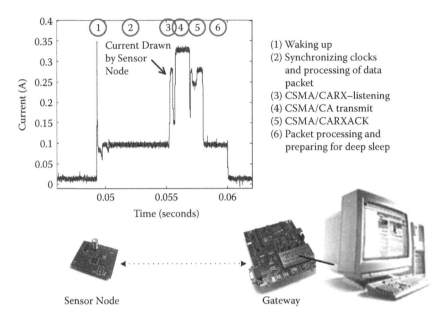

FIGURE 1.12
Current consumption profile of a sensor node communicating with a gateway.

chemical energy directly into electrical energy. They are generally classified into two groups: single-use/primary and rechargeable/secondary batteries. The distinction between the two groups is based on the nature of the chemical reactions.

Primary batteries are discarded when sufficient electrical energy can no longer be obtained from them. Secondary batteries, on the other hand, convert chemical energy into electrical energy by chemical reactions that are essentially reversible. Thus, by passing the electrical current in the reverse direction to that during discharge, the chemicals are restored to their original state, and the batteries are restored to full charge. A supercapacitor (or supercap), is another electrochemical energy system other than batteries that has been gaining its presence in powering wireless sensor nodes. There are several reasons for this phenomenon to occur. One reason is that a supercapacitor is very scalable, and its performance scales well with its size and weight. Another reason is that a supercap has many desirable characteristics that favour the operations of the sensor nodes, such as high power density, rapid charging times, high cycling stability, temperature stability, low equivalent series resistance (ESR), and very low leakage current [23].

Referring to the Ragone plot shown in Figure 1.13, which consolidates various energy storage technologies and compares their power density and energy density characteristics, it can be identified that a supercapacitor has much higher peak power density than the other energy storage device, like batteries and fuel cells. This means that a supercap can deliver more electrical power than batteries and fuel cells within a short time. As shown

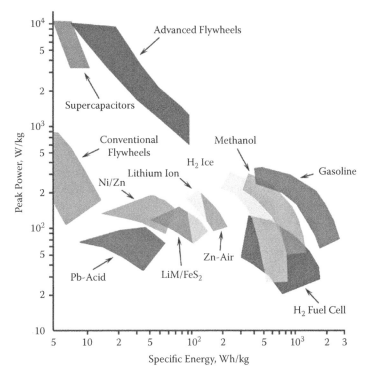

FIGURE 1.13
Ragone plot for comparing the energy storage technologies and their power density versus energy density characteristics. (From J.W. Tester, "Energy transfer and conversion methods," Sustainable Energy Lecture Notes, Topic on Energy Storage Modes, MIT, Cambridge, MA, 2005 [22].)

in Figure 1.13, the peak power densities of supercapacitors are well above 1000 W/kg, whereas the power densities of all types of batteries are in the range of 60 to 200 W/kg, and fuel cells are even lower, which is below 100 W/kg. Hence, for burst power operation, supercapacitors are a better choice than batteries and fuel cells. Conversely, batteries have much higher energy storage capacities than the supercapacitors. This means that batteries can deliver electrical power for a longer period of time as compared to supercaps. Referring to Figure 1.13, it can be seen that the peak energy densities of all types of batteries are in the range of 20 to 200 Wh/kg, whereas the power density of a supercap is below 10 Wh/kg. For sustaining the extended operational lifetime of wireless sensor nodes, solely relying on a supercap might not be suitable due to its very low energy density as compared to the rest of the energy storage devices. Research to increase the energy storage density of both batteries and supercap has been conducted for many years and continues to receive substantial focus [24]. While these technologies promise to extend the lifetime of wireless sensor nodes, they cannot extend their lifetime indefinitely.

Among these nonrenewable energy systems or sources, the rechargeable/alkaline battery is one of the most popular methods for powering the great majority of autonomous sensor nodes in WSNs. The electrical energy necessary for their operation is provided primarily by batteries. Although batteries have been widely used in powering sensor nodes in WSNs presently, the problem is that the energy density of batteries is limited, and they may not be able to sustain the operation of the sensor nodes for a long period of time. Referring to the case scenario of an Xbow sensor node given in Table 1.2 with an operating duty cycle of 1%, the average power consumption of the node with supply voltage of 3 V is around 1 mW. With reference to the highest reported energy for current battery technologies that ranges around 3.78 kJ/cm^3 [1], for the ultralow-power miniaturized wireless sensor node with a volumetric size of around 10 cm^3 operating at an average power consumption of 1 mW to have a 10-year life span, it needs a 100-cm^3 battery. The size of the battery is 10 times the sensor node's size. In fact, the calculation is a very optimistic estimate as the entire capacity of the battery usually cannot be completely used up depending on the voltage drop. In addition, it is also worth mentioning that the sensors and electronic circuits of a wireless sensor node could be far smaller than 10 cm^3. In this case, the battery takes up a significant fraction of the total size and weight of the overall system and also is the most expensive part of the system. Thus, energy supply is largely constrained by the size of the battery.

In short, batteries with a finite energy supply must be optimally used for both processing and communication tasks. The communication task tends to dominate the processing task in terms of energy consumption. Thus, in order to make optimal use of energy, the amount of communication tasks should be minimized as much as possible. In practical real-life applications, the wireless sensor nodes are usually deployed in hostile or unreachable terrains; they cannot be easily retrieved for the purpose of replacing or recharging the batteries, so the lifetime of the network is usually limited. There must be some kind of compromise between the communication and processing tasks in order to balance the duration of the WSN lifetime and the *energy density* of the storage element. In summary, limitation in the device size and energy supply typically means a restricted amount of resources, for example, for central processing unit (CPU) performance, memory, wireless communication bandwidth used for data forwarding, and range allowed. The need to develop an alternative method for powering the wireless sensor and actuator nodes is acute. The main research focus is to resolve the energy supply problems faced by the wireless sensor nodes in a WSN.

1.3 Energy Harvesting Solution for Wireless Sensor Nodes

To overcome the major hindrance of the deploy-and-forget nature of wireless sensor networks due to the limitation of available energy for the network constrained by the high power consumption of the sensor nodes and

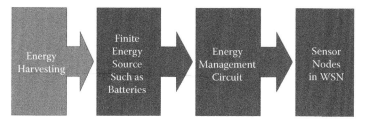

FIGURE 1.14
Paradigm shift from a conventional battery-operated WSN.

the energy capacity and unpredictable lifetime performance of the battery, energy harvesting (EH) technology has emerged as a promising solution for a paradigm shift from a conventional battery-operated WSN to a self-sustainable/autonomous WSN [25–26], as illustrated in Figure 1.14.

1.3.1 Overview of Energy Harvesting

Energy harvesting is a technique that captures, harvests, or scavenges a variety of unused ambient energy sources (e.g., solar, thermal, vibration, and wind) and converts the harvested energy into electrical energy to recharge the batteries. The harvested energy is generally very small (of the order of millijoules) as compared to large-scale EH using renewable energy sources such as solar farms and wind farms, which are on the order of several hundreds of megajoules. Unlike the large-scale power stations that are fixed at a given location, the small-scale energy sources are portable and readily available for usage. Various EH sources, excluding the biological type, that can be converted into electrical energy are shown in Figure 1.15.

Our environment is full of waste and unused ambient energy generated from the energy sources seen in Figure 1.15. These renewable energy sources are ample and readily available in the environment, so it is not necessary to deliberately expend efforts to create these energy sources, like the example of burning nonrenewable fossil fuels to create steam, which in turn would cause the steam turbine to rotate to create electrical energy. Unlike fossil fuels, which are exhaustible, the majority of the environmental energy sources are renewable and sustainable for almost an infinite period. Numerous studies and experiments have been conducted to investigate the levels of energy that could be harvested from the ambient environment. A compilation of various EH sources and their power/energy densities is given in Table 1.3.

Table 1.3 shows the performance of each EH source in terms of the power density factor. It can be clearly observed that there is no unique solution suitable for all environments and applications. According to Table 1.3, it can be observed that a solar energy source yields the highest power density. However, this may not always be the case. Under illuminated indoor conditions, the ambient light energy harvested by the solar panel drops tremendously. The other EH sources could provide higher power density, depending on the

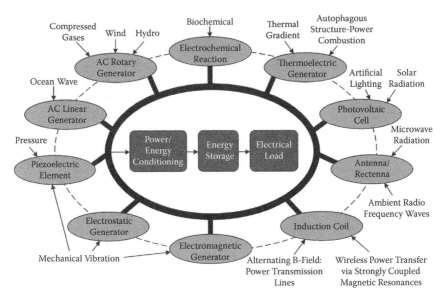

FIGURE 1.15

Energy harvesting sources and their energy harvesters. (Adapted from J.P. Thomas, M.A. Qidwai, and J.C. Kellogg, "Energy scavenging for small-scale unmanned systems," *Journal of Power Sources*, vol.159, pp. 1494–1509, 2006 [27].)

renewable energy sources available at the specific application areas, like an outdoor bright sunny day with a rich amount of solar energy, along a coastal area with a lot of wind energy, on a bridge structure with travelling vehicles and strong vibrations, and so on. In addition, there could also be the possibility of two or more energy sources available for harvesting at the same time. As such, EH technology can provide numerous benefits to the end user, and some of the major benefits of EH suitable for WSNs are stated and elaborated in the following list [27, 28]. An EH solution can do the following:

1. Reduce the dependency on battery power. With the advancement of microelectronics technology, power consumption of the sensor nodes is decreasing, hence, harvested ambient/environmental energy may be sufficient to eliminate the battery completely.

2. Reduce installation cost. Self-powered wireless sensor nodes do not require power cable wiring and conduits; hence, they are very easy to install and reduce the high installation cost.

3. Reduce maintenance cost. EH allows for the sensor nodes to function unattended once deployed and eliminates service visits to replace batteries.

4. Provide sensing and actuation capabilities on a continuous basis in hard-to-access hazardous environments.

TABLE 1.3

Energy Harvesting Opportunities and Demonstrated Capabilities

Energy Source	Performance	Notes
Ambient light	100 mW/cm^2 (direct sunlight)	Common polycrystalline solar cells are 16% to 17% efficient, while standard monocrystalline cells approach 20%
	100 μW/cm^2 (illuminated office)	
Thermal	a. 60 μW/cm^2 at 5 K gradient	Typical efficiency of thermoelectric generators are ≤ 1% for ΔT < 313 K
	b. 135 μW/cm^2 at 10 K gradient	a. Seiko Thermic wristwatch at 5 K body heat
		b. Quoted for a ThermoLife generator at ΔT = 10 K
Blood pressure	0.93 W at 100 mmHg	When coupled with piezoelectric generators, the power that can be generated is on the order of microwatts when loaded continuously and milliwatts when loaded intermittently
Vibration	4 μW/cm^3 (human motion-Hz) 800 μW/cm^3 (machines-kHz)	Predictions for 1-cm^3 generators; highly dependent on excitation (power tends to be proportional to ω, the driving frequency, and y_0, the input displacement)
Hand linear generator	2 mW/cm^3	Shake-driven flashlight of 3 Hz
Push button	50 μJ/N	Quoted at 3 V DC for the MIT Media Lab Device
Heel strike	118 J/cm^3	Per walking step on piezoelectric insole
Ambient wind	1 mW/cm^2	Typical average wind speed of 3 m/s in the ambient environment
Ambient radio frequency	<1 μW/cm^2	Unless near a RF transmitter
Wireless energy transfer	14 mW/cm^2	Separation distance of 2 m

Source: C. Mathna, T. O'Donnell, R.V. Martinez-Catala, J. Rohan, and B. O'Flynn, "Energy scavenging for long-term deployable wireless sensor networks," *Talanta*, vol. 75, no. 3, pp. 613–623, 2008 [29].

5. Provide long-term solutions. A reliable self-powered sensor node will remain functional virtually as long as the ambient energy is available. Self-powered sensor nodes are perfectly suited for long-term applications that are looking at decades of monitoring.

6. Reduce environmental impact. EH can eliminate the need for millions of batteries and energy costs of battery replacements.

Having said that, energy supplied by the EH technique is not perfect but has its own set of challenges:

1. Low power generation densities causes insufficient power for continuous operation

2. Intermittent power supply from energy sources means EH is not feasible in certain situations (e.g., zero solar energy at night)

3. Susceptibility to fluctuating energy sources may affect sensor node operation

4. Expensive in terms of from well to tank as compared to fossil fuels and causes lack of attractiveness

Clearly, it can be deduced from the list of benefits that EH technology is a viable solution to power wireless sensor networks and mobile devices for extended operation with the supplement of the energy storage devices, if not completely eliminating storage devices such as batteries. EH works hand in hand with batteries/supercapacitors to extend devices' operational lifetime.

1.3.2 Energy Harvesting System

In an EH system, there are generally four main components: energy collection and conversion mechanism (energy harvester), electrical power management/conditioning circuit, energy storage device, and electrical load (wireless sensor node) (Figure 1.16). Power output per unit mass or volume (i.e., power/energy density) is a key performance unit for the energy collection and conversion mechanisms. The harvested power must be converted to electricity and conditioned to an appropriate form for either charging the system batteries or powering the connected load directly. Proper load impedance matching between the EH source and the electrical load is necessary to maximize the usage of the harvested energy. Appropriate electronic circuitry for power conditioning and load impedance matching may be available commercially or may require custom design and fabrication.

Referring to Figure 1.16, it can be seen that the function of the energy harvester is to convert energy harnessed from environmental energy sources into electrical energy. Some typical examples of the energy harvesters, as shown in Figure 1.15, include the lead zirconate titanate (PZT) ceramic material that converts mechanical (strain or stress) energy into electrical energy due to the piezoelectric effect, the photovoltaic (PV) cell that converts solar energy into electrical energy, the thermoelectric generator (TEG) output electrical voltage

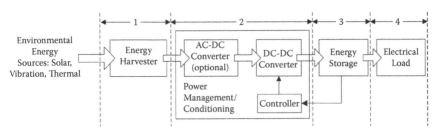

FIGURE 1.16
General block diagram representation of an EH system unit.

when there is a thermal gradient across it, and the wind turbine that converts kinetic energy from wind flow into electrical energy. The harvested electrical energy from the energy harvester needs to be conditioned by some form of power conditioning circuit before supplying it to the load. The main objective of the power electronics technology in the power conditioning circuit as seen in Figure 1.16 is to process and control the flow of electrical energy from the source to the load in such a way that energy is used efficiently. This matching process is a crucial step to ensure that maximum power is transferred from the source to the electrical load. Another function of the power conditioning circuit involves the conversion and regulation of electrical voltage at higher levels into suitable levels for the loads.

To ensure continuity in the load operation even when the external power source is weak or temporarily unavailable, the excess energy already being harvested has to be stored in either the rechargeable battery or supercapacitor as shown in Figure 1.16. Depending on the environmental condition of the ambient energy source, the characteristic of the energy harvester, and the power requirement of the load (i.e., wireless sensor node and control circuitry), each individual EH system is designed and optimized accordingly in order to sustain the operation of the wireless sensor node.

1.3.3 Review of Past Works on Energy Harvesting Systems

There is a significant amount of research works in the literature on harvesting or scavenging small-scale environmental energy for powering wireless sensor nodes. One important point to note is that in order to make the sensor node truly autonomous and self-sustainable in a WSN, the choice of the EH technique is crucial. As such, review of the past works on EH systems is a necessity.

1.3.3.1 *Solar Energy Harvesting System*

The solar energy of an outdoor incident light at midday holds a power density of roughly 100 mW per square centimetre and indicates that in a small volume of 1 cm^2, 100 mW of electrical power can be harvested from the sun by using a solar panel. Conversely, the lighting power density in indoor environments such as illuminated offices drops tremendously to almost 100 μW/cm^2 [30]. Commercially off-the-shelf, single-crystal solar cells offer efficiencies of about 15% and up to 20% to 40% for the state-of-the-art expensive research PV cells recorded by Green et al. in a PV progress report [31]. Thin-film polycrystalline and amorphous silicon solar cells are also commercially available and cost less than single-crystal silicon, but also have lower efficiency of only 10% to 13% [30].

Recently, a number of solar EH prototypes have been presented that perform increasingly efficient energy conversion. Two of the first prototypes were Heliomote [32] and Prometheus [33]. In both systems, the solar panels are directly connected with the storage device. The system Prometheus is depicted in Figure 1.17a. In this case, the solar panel is directly connected to a

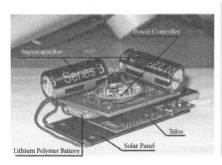

(a) Prometheus sensor node

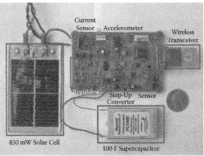

(b) Everlast prototype system

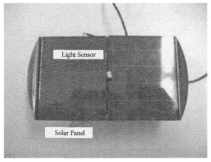

(c) AmbiMax solar panel with light
sensor

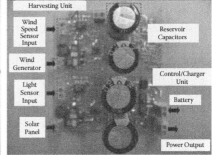

(d) AmbiMax board with supercapacitors

FIGURE 1.17
Examples of a solar EH system. (From X.F. Jiang, J. Polastre, and D.E. Culler, "Perpetual environmentally powered sensor networks," *4th International Symposium on Information Processing in Sensor Networks (IPSN)*, pp. 463–468, 2005 [33]; F.I. Simjee and P.H. Chou, "Efficient charging of supercapacitors for extended lifetime of wireless sensor nodes," *IEEE Transactions on Power Electronics*, vol. 23, no. 3, pp. 1526–1536, 2008 [34]; C. Park and P.H. Chou, "AmbiMax: Autonomous energy harvesting platform for multi-supply wireless sensor nodes," *3rd Annual IEEE Communications Society on Sensor and Ad Hoc Communications and Networks (SECON)*, vol. 1, pp. 168–177, 2006 [35].)

supercapacitor. This means that, especially for low supercapacitor voltages, the solar panel generates much less power than its maximum power P_{MPP}.

Hence, an efficient solar energy harvesting (SEH) system should be able to adapt the electrical operating point of the solar panel to the given light condition so that P_{MPP} is always maintained. For solar panels of a few square centimetres, particular care has to be taken in order not to waste the few milliwatts generated by the solar panel. Everlast [34] is an example that uses the fractional short-circuit current technique (see Figure 1.17b) to achieve maximum power point tracking P_{MPPT}. This technique is easy, simple, and cheap to implement. The voltage V_{MPP} is estimated based on the open-circuit voltage V_{oc} of the solar panel, which is measured periodically by momentarily shutting down the power converter that is connected to the solar panel. However, the drawback with this technique is the transient drop of power during that time where no energy is harvested.

Another SEH system, known as AmbiMax, has been proposed by Park et al. [35]. The AmbiMax system exploits a small photosensor to detect the ambient light conditions and to force the solar panel to work in its maximum power point (MPP) (see Figure 1.17c). Similar to AmbiMax, Dondi et al. presented another circuit [36] that uses a miniaturized photovoltaic module as the pilot panel instead of a photosensor to achieve maximum power point tracking (MPPT) for the SEH system. Indeed, these SEH prototypes have successfully demonstrated that solar energy is a realistic energy source for sensor nodes. However, there is still room for improvement, including system form factor, performance, and so on to suit the power requirement of embedded wireless sensor nodes deployed in application areas such as indoors and in overcast areas where access to direct sunlight is often weak or not available. In some cases where solar energy source may not be a suitable choice, it is required to search for alternative energy sources either to replace the solar energy source as a whole or to supplement the solar energy source when the intensity of the light is low.

1.3.3.2 *Thermal Energy Harvesting System*

Thermal energy is another example of alternative energy sources. Several approaches to convert thermal energy into electricity are currently under investigation (through Seebeck effect, thermocouples, piezo-thermal effect) [37]. The efficiency of these approaches is related to Carnot's law, expressed by $\eta = (T_{max} - T_{min})/T_{max}$. According to Carnot's equation, for a thermal gradient of 5 K with respect to the normal ambient temperature of 300 K, the thermal energy harvesting (TEH) efficiency is computed to be around 1.67%. Consider a silicon device with thermal conductivity of 140 W/mK, as illustrated by Cottone [38], the heat power that flows through conduction along a 1-cm length for $\Delta T = 5$ K is 7 W/cm^2. Hence, the electrical power obtained at Carnot's efficiency is calculated to be 117 mW/cm^2. At first sight, this heat power density of 7 W/cm^2 seems to be an excellent result, but the TEH devices have efficiencies well below the simple Carnot's rule, so the attainable electrical power density turns out to be a small fraction of that, only 117 mW/cm^2. Many research works on TEH devices have been discussed in the literature, and a thermoelectric generator is one of the popular devices that has been developed to harvest thermal energy based on Seebeck's effect. A summary of the implemented TEGs capable of generating from 1 to 60 μW/cm^2 at a 5 K temperature differential is illustrated in a review paper [37] presented by Hudak et al.

A TEH system requires one or more TEGs, heat exchangers on the hot and cold sides of the TEG, a mechanical structure for clamping the heat exchangers to the module and ensuring good thermal contact, thermal insulation to prevent heat losses through the sides, and power electronics for impedance load matching [29]. One commercial application example of TEG is the Seiko Thermic wristwatch, as shown in Figure 1.18a, which is powered by heat generated from the human body.

In the Seiko wristwatch shown in Figure 1.18a, the TEH system consists of a thermoelectric module, a lithium-ion battery, and a simple DC-DC step-up

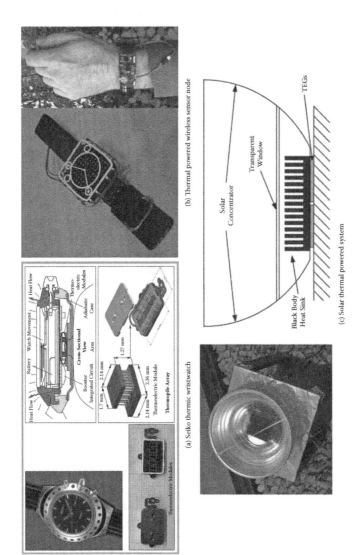

(a) Seiko thermic wristwatch

(b) Thermal powered wireless sensor node

(c) Solar thermal powered system

FIGURE 1.18
Examples of TEH systems. (From N.S. Hudak and G.G. Amatucci, "Small-scale energy harvesting through thermoelectric, vibration, and radio frequency power conversion," *Journal of Applied Physics*, vol. 103, no. 10, pp. 101–301(1-24), 2008 [37]; V. Leonov, T. Torfs, P. Fiorini, and C. Van Hoof, "Thermoelectric converters of human warmth for self-powered wireless sensor nodes," *IEEE Sensors Journal*, vol. 7, no. 5, pp. 650–657, 2007 [39]; H.A. Sodano, G.E. Simmers, R. Dereux, and D.J. Inman, "Recharging batteries using energy harvested from thermal gradients," *Journal of Intelligent Material Systems and Structures*, vol. 18, no. 1, pp. 3–10, 2007 [40].)

voltage regulator [37]. The thermoelectric module of the wristwatch is recorded to yield 60 $\mu W/cm^2$ at a 5 K temperature gradient with 10 TEGs coupled together in series [41]. Similarly, Leonov et al. [39] have considered TEH through thermoelectric power generation from body heat to power wireless sensor nodes as shown in Figure 1.18b. The average power generation at daytime of about 250 μW corresponds to about 20 $\mu W/cm^2$ with a temperature difference of 10 K, which is better than solar panels in many indoor situations, especially considering the TEG power is also available at night. However, these systems do not consider proper matching between the source and the load to ensure MPP operation.

In other TEH research, both Stevens [42] and Lawrence et al. [43] consider the system design aspects for solar-TEH via thermoelectric conversion that exploits the natural temperature difference between the ground and air. Later, Sodano et al. [40] presented a solar-TEH system placed in a greenhouse with a solar concentrator as seen in Figure 1.18c. The solar-TEH system uses a TEG to recharge a NiMH nickel metal hydride battery. At an estimated ΔT of 25 K, the harvested energy was able to recharge an 80-mAh battery in 3.3 min. The authors have demonstrated that a TEG may be used for solar energy conversion as an alternative to photovoltaic devices. However, like before, there are few discussions on the power management aspects of the solar-TEH system.

1.3.3.3 Vibration Energy Harvesting System

The first important virtue of random mechanical vibrations as a potential energy source is that they are present almost everywhere. Mechanical vibrations occur in many environments (e.g., buildings, transports, terrains, human activities, industrial environments, military devices, and so on). Their characteristics are various: spectral shape from low to high frequency and amplitude and time duration manifold depending on the surroundings. Theory and experiments from many research works show that the power density that can be converted from vibrations is about 300 $\mu W/cm^3$ [38]. Devices that convert mechanical motion into electricity can be categorized in electromagnetic, electrostatic, and piezoelectric converters [44, 45]. In the case of electromagnetic converters, a coil oscillates in a static magnetic field and induces a voltage. In electrostatic converters, an electric charge on variable capacitor plates creates a voltage if the plates are moved. Piezoelectric converters finally exploit the ability of some materials like crystals or ceramics to generate an electric potential in response to mechanical stress. A prominent example for the employment of vibrational harvesters is in the watch industry, where vibrational energy converters have been used with success to power wristwatches.

Shenck et al. presented a piezoelectric-powered RFID (radio-frequency identification) system [46] for shoes, as illustrated in Figure 1.19a, that harvests energy from human walking activity. The developed shoe inserts are capable of generating around 10 mW of power under normal walking conditions. This shows that mechanical vibration from human activity is another promising renewable energy source worth investing effort to investigate.

(a) Piezoelectric-powered RFID shoes
with mounted electronics

(c) Electromagnetic vibration-based
microgenerator devices

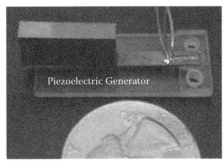

(b) Vibration powered wireless sensor node

FIGURE 1.19

Examples of TEH systems. (From N. Shenck and J. Paradiso, "Energy scavenging with shoe-mounted piezoelectrics," *IEEE Micro*, vol. 21, no. 3, pp. 30–42, 2001 [46]; S.J. Roundy, "Energy scavenging for wireless sensor nodes with a focus on vibration to electricity conversion," Ph.D. Thesis, University of California, Berkeley, 2003 [47]; P. Glynne-Jones, M.J. Tudor, S.P. Beeby, and N.M. White, "An electromagnetic, vibration-powered generator for intelligent sensor systems," *Sensors and Actuators*, vol. 110, no. 1–3, pp. 344–349, 2004 [48].)

A similar approach has been taken by Roundy [47], where piezoelectric generators such as seen in Figure 1.19b have been developed as an attractive method to power wireless transceivers. Other vibration energy harvesting (VEH) research work being reported include wearable electronic textiles [49] and electromagnetic vibration-based microgenerator devices (see Figure 1.19c) for intelligent sensor systems [48]. Meninger et al. have also demonstrated an electromagnetic vibration-to-electricity converter in their research work [50] that can produce 2.5 μW of electrical power in cubic centimetres. Similarly, another piece of research work discussed by Mitcheson et al. [51] indicated that up to 4 μW/cm^3 can be achieved from vibrational microgenerators (on the order of 1 cm^3 in volume) that typical human motion (5 mm motion at 1 Hz) stimulates and up to 800 μW/cm^3 from machine-induced stimuli (2 nm motion at 2.5 kHz). Many meso and microscale EH generators have been developed in the last 5 years [38]; however, there is a lack of adequate ultralow-power management circuit to condition their micropower generation.

In another VEH research work, Paradiso et al. [52] have successfully demonstrated a piezoelectric element with a resonantly matched transformer and conditioning electronics that, when struck by a button, generate 1 mJ at 3 V per 15 N push, enough power to run a digital encoder and a radio that can transmit over 15 m. However, this system requires a large transformer to step up the output voltage generated by the piezoelectric element. The efficiency of the transformer is limited by flux leakage and core saturation when the primary current peaks. Taking an interesting turn, assuming an average blood pressure of 100 mmHg (normal desired blood pressure is 120/80 above atmospheric pressure), a resting heart rate of 60 beats per minute, and a heart stroke volume of 70 ml passing through the aorta per beat [53], then the power generated is about 0.93 W. Ramsay et al. [54] found that when the blood pressure is exposed to a piezoelectric generator, the generator can generate power on the order of microwatts when the load applied changes continuously and milliwatts as the load applied changes intermittently. However, harnessing power from blood pressure would only limit the application domains to wearable microsensors.

1.3.3.4 Wind Energy Harvesting System

Like any of the commonly available renewable energy sources, wind energy harvesting (WEH) has been widely researched for high-power applications where large wind turbine generators (WTGs) are used for supplying power to remote loads and grid-connected applications [55, 56]. According to a study by the National Renewable Energy Laboratory (NREL) [57], wind energy is the fastest-growing electricity-generating technology in the world. In the past 10 years, global installations of wind energy systems have grown at least 10-fold, from a total capacity of 2.8 gigawatts (GW) in 1993 to almost 40 GW at the close of 2003 [58]. In spite of this continuing success of WEH at a large scale, there have been very few attempts on the development of small-scale WEH, those that are miniaturized in size and highly portable, to power small autonomous sensors deployed in remote locations for sensing or even to endure long-term exposure to a hostile environment such as a forest fire. Although very few research works are reported in the literature on small-scale WEHs, some efforts to generate power at a very small scale have been made recently. Park et al. presented a MPPT for a small windmill [35]. The WEH system, as illustrated in Figure 1.20a, exploits the near-linear relationship between the wind speed and the rotating frequency of the WTG's rotor to force the WTG to work in its MPPs.

Similarly, a small-scale WTG has been presented by Holmes et al. [59], and the power densities harvested from air velocity are quite promising. Later,

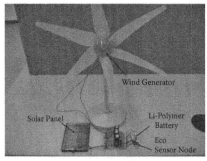

(a) AmbiMax hardware with a solar panel,
wind generator, lithium polymer battery,
and Eco Node

(b) Piezoelectric windmill prototype

(c) Optimized design of the small-scale windmill with an isometric view
and internal crankshaft structure exhibiting the translation mechanism

FIGURE 1.20
Examples of TEH systems.

Weimer et al. [60] presented an anemometer-based solution to perform the WEH and sensing tasks, which are accomplished separately by two different devices. The authors utilize the motion of the anemometer shaft to turn a coupled alternator to generate electrical power for the sensor nodes. Although the sensor nodes incorporating the harvesting solution have an increased operational lifetime, this comes with the price of larger device size, higher overall cost, and higher energy conversion loss.

In another WEH research work [61], Priya et al. designed a windmill that uses piezoelectric elements to generate electricity from wind energy (see Figure 1.20b). An output power of 10.3 mW was harvested and reported for a wind flow that leads to 6 rotations per minute. Subsequently, another group of researchers, Myers et al. [62], developed an optimized small-scale piezoelectric windmill as shown in Figure 1.20c. The whole structure of the windmill is made of plastic, and it utilizes 18 piezoelectric bimorphs to convert wind energy, hence vibration energy, into electrical energy. The windmill was tested to provide 5 mW of continuous power at an average wind speed of

4.5 m/s. Still, the physical size of the WEH systems are too large and bulky as compared to the sensor node, and their harvested power exceeds the power requirement of the sensor node. In addition, there is a lack of adequate power management circuit to maximize electrical power transfer from the source to the load.

1.4 Contribution of This Book

As described in the literature survey on EH systems, there is no definite energy source that is suitable for sustaining the operation of a wireless sensor node in a different variety of applications. Hence, this book aims to tell how to go about the design, followed by some analysis for further optimization, and practical implementation of the various types of EH autonomous sensor systems. In many WSN applications, the environmental conditions of the sensed area, where the sensor nodes are deployed, are often not consistent. The energy sources are intermittent and fluctuating in nature, while the operations of the wireless sensor nodes require a constant electrical power source. As such, one of the main focuses of the book is the adequate power management circuit design to provide a proper match between the EH mechanism and the sensor node. Other key contributions of the research works recorded in this book are as follows:

- Two types of small-scale WEH systems have been proposed: (1) direct WEH approach using a WTG and (2) indirect WEH approach using piezoelectric material. Detailed analysis and characterisation of the wind energy conversion mechanisms have been provided. Based on the characteristics of the WEH mechanisms obtained, the power management unit is designed to take care of the dynamics of both the WEH mechanisms subjected to environmental condition variation, such as varying wind speeds and the wireless sensor node operating in its WSN. Since most of the conventional MPPT algorithms are not suitable for the WTG, a resistor emulation or impedance-matching scheme has been introduced for MPPT. In addition, an alternating current–direct current (AC-DC) active rectifier has been designed using MOSFETs (metal-oxide-semiconductor field-effect transitors) in place of diodes for rectifying the low-amplitude AC voltage generated by the WTG under low wind conditions. Prototypes of the WEH systems have been developed to validate the performance of the systems.
- In some places, a wind energy source might not be available; hence, TEH has been investigated. The TEH mechanism, which houses a miniaturized TEG, has been designed to harvest the thermal energy from the heat source. An equivalent electrical circuit model of the TEH mechanism has been developed. Based on this equivalent model, the

thermal and electrical analyzes of the TEH mechanism are illustrated. Since the characteristic of the TEH mechanism is found to be similar to the WTG, a resistor emulation-based MPPT is developed to naturally track the MPP of the TEH mechanism with very little control circuit.

- A batteryless and wireless remote controller has been developed to switch electrical appliances such as lights and fans on/off in a wireless manner. Two types of piezoelectric-based VEH systems are presented to harvest impact or impulse forces from a human pressing a button or switch action. Detailed understanding and characterisation of the performance of the VEH mechanisms are carried out. Since the harvested power is lower than the power consumed by the wireless RF transmitter, an energy management circuit has been implemented. The harvested energy from the VEH mechanism is first accumulated and stored in a capacitor until there is sufficient stored energy to power the RF transmitter; the transmitter is then energized.

- For some WSN applications, multiple energy sources are available. A hybrid wind and SEH scheme is proposed to extend the lifetime of the wireless sensor node. In order for the developed hybrid energy harvesting (HEH) system to harvest simultaneously from both energy sources, the WEH subsystem uses the resistor emulation technique while the SEH subsystem uses the constant voltage technique for MPP operation. In another HEH research work, a hybrid of indoor ambient light and TEH has been proposed. Energy sources of different characteristics are connected directly. A detailed analysis of their relationship has been conducted to validate that the impedance mismatch issue does not affect much of the maximum attainable power from the HEH mechanism. A power management circuit is developed to condition the combined output power harvested from both energy sources.

- Like EH, two remote means of charging low-power electronic devices are proposed. A magnetic inductive approach has been investigated to transfer electrical power from the power lines to the sensor node in a wireless manner. Based on the electrical power requirements of the wireless RF transmitter and the electrical characteristic of the magnetic energy harvester, a magnetic induction system has been designed and successfully implemented. Another wireless power transfer (WPT) mechanism operating at its magnetic resonance has been proposed to further extend the distance for wireless power transmission. Detailed theoretical analysis of the WPT mechanism has been provided and then verified by the simulation results. Based on the verified theoretical findings and experimental results, the design of the WPT mechanism can be optimized for improving the WPT distance, efficiency, and form factor.

- Various types of EH systems and their respective main components, namely, energy harvester (source), power management circuit, energy storage device, and wireless sensor node (load), have been proposed, investigated, and analyzed. These EH systems have been designed and optimized to suit the target applications (i.e., ambient conditions and event/task requirements) and then implemented into hardware prototypes for proof of concept.

1.5 Organization of the Book

Chapter 1 has introduced the background of the research works recorded in the book. Motivation for this research has been stated. The problems in powering the wireless sensor nodes have been identified. To overcome these problems, an overview of the EH solution and its system design were described. A brief review of past works on various EH systems has been provided to show the state of the art. Contributions of the research works recorded in the book have been listed.

Chapter 2 discusses WEH research. Two types of WEH approaches based on WTG and piezoelectric wind energy harvester have been explored. For the first approach, the issues of small-scale WEH using a WTG for sustaining the operation of a wireless sensor node are discussed. To resolve the problems, an ultralow-power management circuit consisting of a resistance emulator and an active rectifier is specially designed to optimize the WEH system. Detailed analysis of the WEH system is provided and then validated by the experimental results. For space constraint applications, a conventional WTG is not suitable. In the second part of Chapter 2, a novel method to harvest wind energy using piezoelectric material (PZT) is proposed. Energy harvested from the piezoelectric-based wind energy harvester is first accumulated and stored in a capacitor until sufficient energy is harvested; a trigger signal is then initiated to release the stored energy in the capacitor to power an autonomous wind speed sensor node. Experimental results are provided to verify the novel method proposed in this work.

Chapter 3 discusses TEH from ambient heat sources with low temperature differences. TEH has recently received great attention but is impeded by the challenges of low energy conversion efficiency, inconsistency, and low output power due to temperature fluctuation and higher cost. To supplement the TEH scheme in sustaining the operation of a wireless sensor node, this chapter presents a DC-DC buck converter with resistor emulation-based MPPT. The resistance emulator approach uses a specially designed ultralow-power management circuit to perform close impedance matching between the thermal energy harvester and the sensor node to achieve MPPT under varying thermal conditions. Detailed design steps are provided for obtaining various parameters of the resistance emulator used in this work. Experimental results validate the performance of the optimized TEH system.

Like any of the commonly available renewable energy sources, vibration is another type of energy source. Human activity can be the source of vibrational energy. Chapter 4 presents two types of piezoelectric-based VEH systems to harvest impact or impulse forces from a human pressing a button or switch action. The issues with the conventional approach of using electrical cables in residential and industrial buildings to connect the appliance to the control switch on the wall have been a cause of nuisance. To resolve the problem, a batteryless and wireless remote controller has been developed to switch electrical appliances such as lights and fans on/off in a wireless manner. The experimental results verify that by depressing (1) the piezoelectric push-button igniter or (2) the prestressed piezoelectric diaphragm material, electrical energy is generated and stored in the capacitor. Once sufficient energy is harvested, the batteryless and wireless remote controller is powered up for operation.

An EH system itself has an inherent problem: the intermittent nature of the ambient energy source. Hence, the operational reliability of the wireless sensor node may be compromised. To augment the reliability of the wireless sensor node's operation, Chapter 5 discusses two HEH approaches. A hybrid WEH and SEH scheme is proposed to harvest simultaneously from both energy sources in order to extend the lifetime of the wireless sensor node. When the two energy sources with different characteristics are combined, it is bound to have the issue of impedance mismatch between the two different sources and the load. To overcome the problem, each energy source has its own power management unit to maintain at its respective MPPs. Experimental results obtained show that the electrical power harvested by the optimized HEH system is much higher than the single-source-based WEH. Chapter 5 also presents a hybrid of indoor ambient light and TEH scheme that uses only one power management circuit to condition the combined output power harvested from both energy sources. An efficient microcontroller-based ultralow-power management circuit with fixed voltage reference-based MPPT is implemented with a closed-loop voltage feedback control to ensure near maximum power transfer from the two energy sources to its connected electronic load over a wide range of operating conditions. Experimental results are provided to validate the proposed HEH scheme by directly connecting the two energy sources in a parallel configuration.

Other than EH, Chapter 6 also demonstrates an alternative means to remotely power low-power electronic devices through WPT mechanisms. The WPT mechanism uses the concept of inductive coupling (i.e., by harvesting the stray magnetic energy in power lines to transfer electrical power without any electrical connection). Experimental results obtained validate the performance of the developed WPT system. To extend the WPT distance, self-resonating coils, operating in a strongly coupled mode, are demonstrated. Detailed theoretical analysis of the WPT system are provided and then verified by the simulation results. Experimental validation of the performance of the WPT system is provided.

Chapter 7 concludes the book. It briefly states the focus areas and then discusses the proposed solutions. It also lists the possible future work in this line of research.

1.6 Summary

This chapter introduces the problem of a battery with a limited energy supply, hence an operational lifetime of wireless sensor nodes, and explains the motivation for the research works recorded in the book. Basic EH and its system design are described. A literature survey of the past work in this area is provided. The main contributions of this book are then listed. The structure of this book is provided along with the focus area of each chapter. The next chapter elaborates on the small-scale WEH research.

2

Wind Energy Harvesting System

Like any of the common renewable energy sources, wind energy harvesting (WEH) has been widely studied and researched for high-power applications where large wind turbine generators are used for supplying power to remote loads and grid-connected applications [55–56]. However, to my best knowledge, few research works can be found in the literature that discuss the issue of small-scale WEH [35, 60], and those are miniature in size and highly portable to power small autonomous sensors deployed in remote locations for sensing or even to endure long-term exposure to a hostile environment such as a forest fire. In addition, energy harvesting from wind, the incoming wind speed can be sensed to determine the outbreak of a disaster and raise an emergency warning to people in that area so that they have time to respond to the emergency situation. Thus, this reduces the number of causalities caused by the disaster.

In this chapter, two types of small-scale WEH schemes are introduced: (1) the direct WEH approach using a wind turbine generator in Section 2.1 and (2) the indirect WEH approach using piezoelectric material in Section 2.2. Several challenges are associated with these types of small-scale WEH systems in contrast to the large-scale WEH systems, such as:

1. Random wind at the concealed deployment site of small-scale WEH, rather than regular wind flow in open field, results in an extremely intermittent and fluctuating wind energy supply.

2. Low wind speeds at low height and concealed ground, in the range of 2 to 7 m/s, as compared to large-scale WEH systems with tall towers up to 90 m, which are exposed to high wind speeds of greater than 10 m/s, significantly reduce the harvested power by a cube root factor.

3. Weak aerodynamic force is generated with a very small wind front area of few square inches in comparison to a huge wind turbine, which further lessens the available wind power for harvesting.

4. A small size, lightweight, and low-cost WEH system is required by miniaturized ubiquitous wireless sensor nodes instead of a large-scale WEH system for power generation.

Due to the challenges, the proposed direct and indirect WEH systems have to be lightweight and small in size, comparable to the miniaturized wireless sensor node, and able to harvest sufficient energy from weak and uncertain wind energy sources to sustain the operation of the sensor node for a long period of time.

2.1 Direct Wind Energy Harvesting (WEH) Approach Using a Wind Turbine Generator

To better understand the functionality and performance of the direct WEH system, the circuit architecture of a wind-powered wireless sensor node is presented in Figure 2.1. The wind-powered sensor node in Figure 2.1 consists of three main building blocks: (1) an energy harvester incorporating the wind turbine coupled to the electrical generator; (2) a power management unit, which consists of a power conditioning circuit and energy storage, and (3) the wireless sensor node itself.

For a space-efficient WEH system operating at low wind speeds, the AC (alternating current) voltage V_p (peak), generated by the wind turbine generator is in the range of 1 to 3 V, which is relatively smaller than that of the large wind turbine with a megawatt power rating and an output voltage of hundreds of volts [55]. It is thus challenging for the AC-DC (direct current) rectifier using conventional silicon-based diodes, which has a high on-state voltage drop V_{on} of 0.7 to 1 V, to rectify and convert the low-amplitude AC voltage into a form usable by the electronic circuits. Based on the instantaneous output DC voltage given by $V_{dc} = V_p - 2 * V_{on}$, it can be seen that the efficiency of the AC-DC rectifier using diodes is especially low for the small-scale WEH system. To overcome this challenge, metal-oxide-semiconductor field-effect transistors (MOSFETs) are used in place of the conventional diodes because the voltage drop across a MOSFET is smaller than that with a diode. A MOSFET-based active rectifier using comparators to sense the zero crossings of the input AC voltage has been proposed in the literature [63–65] to achieve high AC-DC power conversion efficiency in very-low-voltage applications. The problem

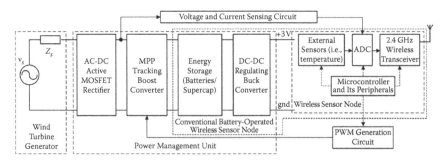

FIGURE 2.1
Functional block diagram of a wind energy harvesting (WEH) wireless sensor node.

with this approach of sensing voltage is that if the input voltages of each of the comparators are too close to each other, excessive oscillations occur, and the AC-DC rectifier's efficiency is drastically reduced. To enhance the performance of this active rectifier, a current sensing approach for generating the gating signals for the MOSFET-based active rectifier is designed and implemented in the chapter.

Another challenging issue addressed in this chapter is that the electrical power harvested by the WEH system for powering the wireless sensor nodes is often very low, on the order of the milliwatt range or less. The situation is even worse if the wind turbine generator is not operating at its maximum power point (MPP). The primary concern is to develop a high-efficiency power converter using micropower-associated electronic circuits to track and maintain maximum output power from the wind turbine generator to sustain the wireless sensor node operation over a wide range of operating conditions. Maximum power point tracking (MPPT) techniques have been commonly used in large-scale WEH systems [66–68] for harvesting a much higher amount of energy from the environment. However, these MPPT techniques require high computational power to fulfill their objective of precise and accurate MPPT. Implementation of such accurate MPPT techniques for a small-scale WEH whereby the power consumed by the complex MPPT circuitry is much higher than the harvested power itself therefore is not desirable. So far, very limited research works can be found in the literature that discuss a simple but compatible MPPT algorithm addressing the issue of a small-scale WEH system. In this chapter, the resistor emulation approach is investigated for a micro wind turbine generator. The rationale behind the resistor emulation approach is that the effective load resistance is controlled to emulate the internal source resistance of the wind turbine generator [69–71] to achieve good impedance matching between the source and load, and hence the harvested power is always at its maximum at any operating wind speed.

The emphasis of this chapter is on resolving the two mentioned challenges associated with a small-scale WEH wireless sensor node using the designed ultralow-power management circuit with little overhead power consumed. The rest of the chapter is organized as follows: Section 2.1.1 describes the details of the wind energy conversion system and determines the output power available at each stage of the system. Section 2.1.2 discusses the issues related to the design of an efficient power management circuit to interface the wind turbine generator and the wireless sensor node. Following that, the experimental results of the optimized WEH wireless sensor node prototype are illustrated in Section 2.1.3, with the conclusion reported in Section 2.1.4.

2.1.1 Wind Turbine Generators

Within the WEH system, the energy harvester, which converts the raw wind energy harnessed from the ambient environment into AC or DC electrical energy depending on the type of generator used, is first investigated. In this case, the energy harvester is a horizontal-axis micro wind turbine with a

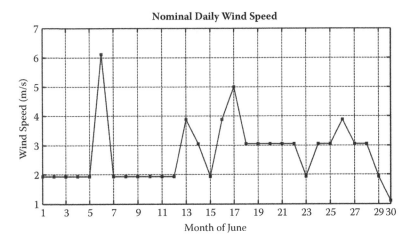

FIGURE 2.2
Nominal monthly wind speed in a typical deployment location over a period of 30 days in 2006.

blade radius of 3 cm directly coupled to a single-phase AC generator with a volumetric size of 1 cm^3. The process starts with quantifying the wind power P_{wind} using the relationship between the input wind speed v in metres/second and the output power available in wind P_{wind} in watts expressed as [72]

$$P_{wind} = \frac{1}{2}\rho A v^3 \qquad (2.1)$$

where A is the given wind front contact area in square metres. Using the nominal daily wind speed recorded throughout the month in a sample remote area environment as shown in Figure 2.2 [73], the average wind speed throughout the month is calculated to be around 3.62 m/s. Based on the Beaufort scale, the calculated average wind speed is termed as fairly light, which is equivalent to a gentle breeze, and the amount of energy available in the wind for harnessing is therefore quite limited. Under gentle breeze conditions, the power available from the wind is computed based on Equation 2.1 to be 82 mW. As can be seen in Figure 2.2, the fluctuation in wind speed is quite significant; wind speed can go as high as 6 to 7 m/s and resides at a low wind speed of around 2 m/s for many days. This high fluctuation in wind speed, due to the geographical condition of the sample remote area, requires the power management circuit to have a wide input operating bandwidth to be able to cover the minimum as well as maximum harvested electrical power.

Figure 2.3 is a functional block diagram of the wind turbine generator system showing the conversion of raw wind power to electrical power, indicating the corresponding power available at different stages, for example, the raw wind power P_{wind}, the mechanical power P_T, and the electrical power P_{elec}. When wind flows past the blades of the wind turbine, some part of

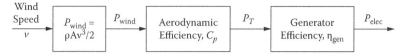

FIGURE 2.3
Functional block diagram of a wind turbine generator.

the power available in the wind is harvested by the wind turbine to generate electricity. It is experimentally tested that the aerodynamic efficiency (also known as the power coefficient C_p) of the wind turbine with a wind speed of 3.62 m/s is around 39% ($C_p = 0.39$). Using all the technical information gathered for the wind turbine generator, the theoretical power P_T generated by the wind turbine at an average wind speed of 3.62 m/s can be calculated as

$$P_{T,3.62 \text{ m/s}} = C_p P_{wind} \qquad (2.2)$$
$$= \frac{1}{2}(0.39)(1.225)[\pi(0.03^2)](3.62^3) = 32 \text{ mW}$$

The estimated power of 32 mW is the mechanical power available at the rotor shaft of the wind turbine based on the aerodynamic blade effect as well as the rotor shaft of the direct-coupled electric generator. Referring to Figure 2.3, taking the generator efficiency η_{gen} into the power flow calculation, the amount of electrical power harvested by the wind turbine P_{elec} can be described in terms of the incoming wind speed by the following equation:

$$P_{elec,3.62 \text{ m/s}} = \eta_{gen} P_{T,3.62 \text{ m/s}} \qquad (2.3)$$
$$= (0.41)(32 \text{ mW}) = 13.12 \text{ mW}$$

The electrical powers P_{elec} obtained experimentally from the wind turbine over a range of simulated wind speeds of 1.3 to 8.5 m/s were tested with different resistance loadings, and the power curves of the wind turbine are shown in Figure 2.4. For wind speeds available at the deployment site ranging from 2.3 to 7 m/s as shown in Figure 2.2, Figure 2.4 shows that the harvested electrical power ranges from 2 to 70 mW, respectively. For the average wind speed at the target site of 3.62 m/s, the electrical average output power generated by the wind turbine is 13 mW. This maximum electrical power of 13 mW can only be harvested from the wind turbine at a matching load resistance of 100 Ω.

2.1.2 Design of an Efficient Power Management Circuit

In the WEH system, the power management unit is used to take care of the dynamics of both the wind turbine generator subjected to environmental condition variations such as varying wind speeds as well as the power-aware wireless sensor node operating at the deployment site. Inside the power management unit, a supercapacitor, which has a fast dynamic response, is

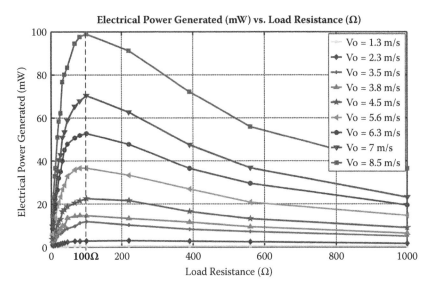

FIGURE 2.4
Power curves of a wind turbine generator over a range of load resistances.

employed to decouple the interrelationship between the fluctuating energy supply of the wind turbine generator and the duty-cycling operation of the wireless sensor node. It is necessary to optimize the power management unit with consideration of the characteristic of the wind turbine generator and the performance of the sensor node in order to meet the application requirement. As such, a WEH system optimized using a specially designed ultralow-power management circuit with two distinct highlights has been presented: (1) an AC-DC active rectifier using MOSFETs in place of diodes for rectifying the low-amplitude AC voltage generated by the wind turbine generator under a low wind condition and (2) a DC-DC boost converter with a resistor emulation algorithm to perform MPPT.

2.1.2.1 Active AC-DC Converter

The active AC-DC converter can be separated into two stages: the negative voltage converter and the active diode. The first stage of the active rectifier circuit, as seen in Figure 2.5, is made up of four standard MOSFETs: two high-side P-type MOSFETs, PMOS1 and PMOS2, employed to deliver the positive and negative half cycles v_1 and v_2, respectively, of the AC voltage to the output DC voltage V_{dc}; and two low-side N-type MOSFETs (NMOS1 and NMOS2) that provide a path for the ground node V_{gnd} to return to the lower potential of either v_2 or v_1, respectively. By doing so, the first stage of the active rectifier converts the negative half wave of the input sinusoidal wave into a positive one. Moreover, no additional start-up circuit is

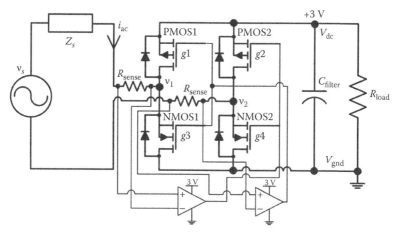

FIGURE 2.5
Schematic diagram of a full-bridge active MOSFET rectifier.

necessary in this configuration because the current has an alternative path through the body diode of the MOSFET whenever the MOSFET switch is not activated.

The main function of the second stage of the rectifier is to control the on and off switching sequences of the MOSFETs residing in the first stage of the active rectifier to facilitate the rectification process. According to Lam et al. [64] and Seeman et al. [65], the proposed active diode stage uses a fast and low-power comparator circuit that continuously samples the voltage across each of the MOSFET switches. The problem with this kind of voltage-sensing approach is that if the voltage difference between the input AC voltage v_1 or v_2 and the output DC voltage V_{dc} is small or the comparator is too slow, then there are excessive oscillations occurring during the on and off switching transitions as illustrated in the zoomed waveforms in Figure 2.6. Due to these unwanted oscillations, some part of the harvested electrical energy is lost, resulting in the reduction in the overall efficiency.

To overcome the problem associated with very small sensing voltage across the MOSFET switches g_1 to g_4 as shown in Figure 2.5, an alternative approach is to sense the AC source current i_{ac} as shown in Figure 2.5 to generate the respective gating signals for the MOSFET switches g_1 to g_4 in the active rectifier circuit. The proposed current-sensing circuit is specially designed to be simple and power saving using two precise zero-crossing comparators (LMC7215) with very low overhead power consumption of 90 µW. The source voltage of the comparators of 3 V is taken from the storage capacitor as shown in Figure 2.5.

Whenever the peak value of the input AC voltage during either the positive or negative half-cycles v_1 or v_2, respectively, is higher than the output

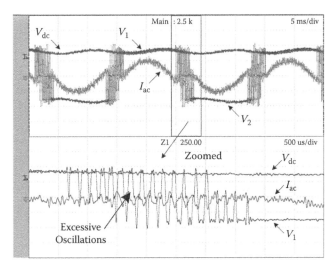

FIGURE 2.6
Experimental waveforms of an active MOSFET full-bridge rectifier using a voltage sense approach resulting in excessive oscillations.

DC voltage V_{dc}, current i_{ac} flows through the sensing resistor R_{sense}, and the zero-crossing comparators sense the flow of the AC current by measuring the voltage across R_{sense}. The comparators generate the control signals to turn the respective combinations of the MOSFET switches on or off, that is, (1) positive half AC cycle: PMOS1 and NMOS2 on, PMOS2 and NMOS1 off; and (2) negative half AC cycle: PMOS2 and NMOS1 on, PMOS1 and NMOS2 off. During the switching transition of the MOSFET switches between on and off states, it can be clearly observed from the zoomed waveforms of Figures 2.6 and 2.7 that using the voltage sense approach introduces more unwanted oscillations (Figure 2.6) than using the current sense approach (Figure 2.7). The use of the voltage sense approach leads to more energy loss incurred in the active rectifier, thus reducing its overall power conversion efficiency. Hence, the current sense approach, which yields better performance than its counterpart, is employed in the active rectifier to control the switching of the MOSFET switches in the bridge rectifier bridge.

The performance of the MOSFET-based active rectifier using the proposed current sense approach is compared with the conventional diode-based passive rectifier to seek a better solution for rectifying low-amplitude AC voltages generated by a micro wind turbine, especially at low wind speeds. The performance comparison is based on the voltage drops measured across the two diodes and the two MOSFETs of the rectifiers, which are plotted over a range of the load resistances as shown in Figure 2.8. At optimal load of 150 Ω, it is observed that the voltage drop in the passive rectifier at a wind speed of 3.62 m/s is around 0.6 V, which is four times more than the voltage loss in it of around 0.15 V. With less voltage drop in the active rectifier, the active rectifier, it is able to rectify much lower amplitude input AC voltages than

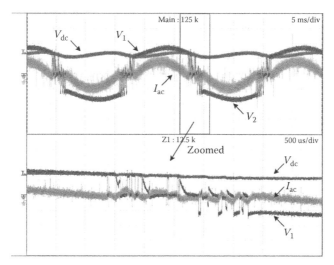

FIGURE 2.7
Experimental waveforms of an active MOSFET full-bridge rectifier using the proposed current sense approach at an optimal load condition.

the passive rectifier, which allows the wind turbine to continue harvesting even for very low incoming wind speeds. In addition, more electrical power is harvested from the wind turbine (see Figure 2.9), thus, a higher AC-DC power conversion efficiency is achieved over a wide range of wind speeds as shown in Figure 2.10.

Consolidating the experimental data collected for both the passive and active rectifiers, the efficiencies of both rectifiers are calculated using

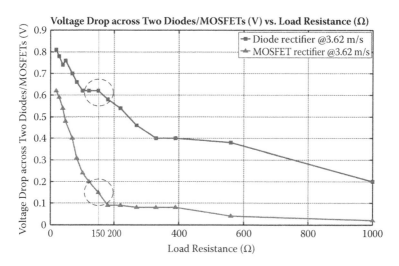

FIGURE 2.8
Voltage drop comparison between diodes and MOSFETs (V) over a range of load resistance (Ω).

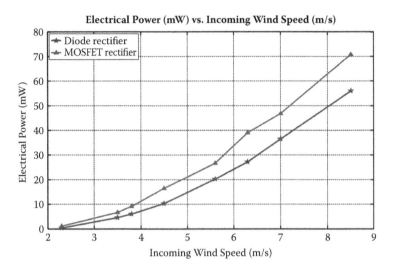

FIGURE 2.9
Electrical power generated by active and passive rectifiers at optimal load conditions.

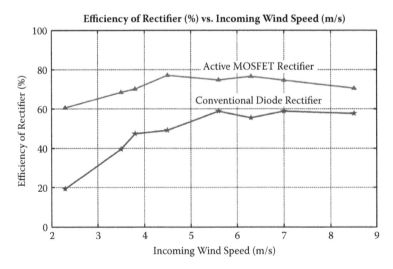

FIGURE 2.10
Efficiency comparison between active and passive rectifiers under optimal load conditions.

Equation 2.4 (under unity power factor operation as illustrated in Figure 2.7) and then plotted in Figure 2.10 for comparison.

$$\eta_{rect} = \frac{P_{dc}}{P_{ac}} * 100\% = \frac{V_{dc,meas}^2 / R_{load}}{V_{ac,meas} * I_{ac,meas} * (\cos\phi \simeq 1)} * 100\% \qquad (2.4)$$

Over the span of wind speeds from 2.3 to 8.5 m/s, the efficiency of the MOSFET-based active rectifier is on average 15% to 25% higher than the diode-based passive rectifier (see Figure 2.10). This improvement in the efficiency of the AC-DC conversion process is mainly due to the replacement of diodes with very low on-state voltage drop MOSFETs and its associated ultralow-power current sensing and control circuit. It is worth noting that the power loss incurred in the current sensing and control circuit is only 90 μW, which is a small fraction of the total harvested power as seen in Figure 2.9, so it does not pose any significant electrical loading to the main WEH system. In addition, even though the components used in active rectifiers such as MOSFETs and operational amplifiers are more expensive than a simple diode rectifier, the surplus in harvested power is very crucial in a small-scale WEH system. Hence, the proposed active rectifier holds great importance in the design of the power management unit of the WEH system.

2.1.2.2 Boost Converter with Resistor Emulation-Based Maximum Power Point Tracking (MPPT)

Unlike standard voltage-regulating boost converters, the main functions of the boost converter in the power management unit of the WEH system are (1) to step up the low DC voltage output of the wind turbine V_{dc} to charge the energy storage device and (2) to perform MPPT so that maximum power transfer takes place. Depending on the energy storage level of the supercapacitor, the output voltage of the wind generator V_{dc} is manipulated to transfer maximum power to the supercapacitor by adjusting the duty cycle of the pulse width modulation (PWM) gate signal of the boost converter such that V_{dc} is as close as possible to V_{mppt}, the voltage at which the harvested power is at its maximum.

There are different algorithms proposed to date for seeking the MPP for stand-alone photovoltaic systems [74] as well as for large-scale wind turbines [66]. According to Salas et al. [74], the MPPT algorithms can be grouped into direct and indirect methods. The direct method involves an iterative and oscillating search for the MPPs, resulting in excessive energy loss during the search process that is very undesirable for small-scale WEH systems. With respect to the commonly used indirect methods, refer to the electrical power harvested curves against the generated voltage and current as shown in Figures 2.11 and 2.12, respectively. It is observed that the MPPT voltage (V_{mppt}) and current (I_{mppt}) at which the harvested power is maximum are different for different incoming wind speeds; hence, variables such as V_{mppt} and I_{mppt} cannot be set as references for the indirect MPPT method.

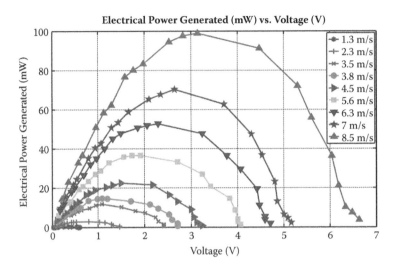

FIGURE 2.11
Power harvested by a wind turbine generator plotted against generated voltage for a range of
wind speeds.

Since most of the conventional MPPT algorithms are not suitable for the
WEH system, this chapter presents an alternative MPPT technique based on
the concept of emulating the load impedance to match the source impedance
as described by Paing et al. [70] and Erickson and Maksimovic [71]. This
technique is also known as resistor emulation or impedance matching. The
power curve plotted in Figure 2.4 shows that when the load resistance matches
the source resistance of the wind turbine generator, the harvested power is

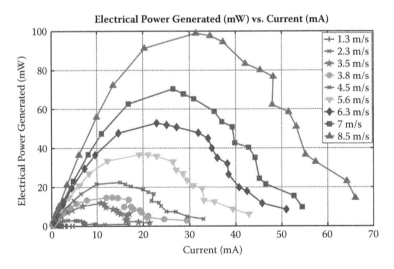

FIGURE 2.12
Power harvested by a wind turbine generator plotted against generated current for a range of
wind speeds.

always at a maximum for different wind speeds. However, for other loading conditions, shifting away from the internal resistance of the wind turbine generator, either very light or heavy electrical loads, the electrical output power being generated by the generator drops significantly. This exhibits that the MPPT technique based on resistor emulation is a possible option to assist the small-scale WEH system to achieve maximum power harvesting from the wind turbine generator.

Khouzam and Khouzam [69] discussed the direct-coupling approach for optimum load matching between the energy harvester and its load by carefully selecting the harvester's rated parameters with respect to the load parameters. Another resistor emulation approach proposed by Paing et al. [70] is to operate the boost converter as an open-loop resistor emulator with proper selection of the components to naturally track the MPP to match the optimal load impedance for the energy harvester. Both approaches require some form of initial tuning as well as the load impedance needs to be fixed. However, in practice, this may not be the case as the load impedance tends to change just like the charging and discharging process of the supercapacitor; therefore, the direct-coupling method as well as the simple open-loop resistor emulation method may not be suitable in this context. To overcome that, a microcontroller-based resistance emulator with a closed-loop feedback resistance control scheme is proposed as the MPP tracker of the WEH wireless sensor node for various dynamic conditions. The proposed scheme does not require any initial tuning, unlike those two existing approaches, because there is a microcontroller, together with its feedback resistance, to automatically tune the WEH system to its MPPT points. In addition, the proposed MPP tracker is embedded with the closed-loop control feature to continuously track and emulate the reference optimal resistance as the load impedance changes. The designed boost converter circuitry with the resistor emulation MPPT approach is depicted in Figure 2.13. It is essentially composed of three main building blocks: (1) a boost converter to manage the power transfer from the wind turbine to the load (i.e., power management unit, supercapacitor, and wireless sensor network [WSN]); (2) an MPP tracker based on a resistor emulation approach and its sensing and control circuit that manipulates the operating point of the wind turbine to keep harvesting power at the maximum power point; and (3) a PWM generation circuit. Using the voltage and current-sensing circuit, the feedback resistance signal R_{fb} is obtained and compared with the reference resistance signal $R_{opt,ref}$ in a microcontroller to perform the closed-loop MPPT control of the boost converter via the PWM generation circuit. The PWM generation circuit is used to multiply the low-frequency PWM control signal (<100 Hz) generated from the low-power microcontroller to a much higher switching frequency (10 kHz) so that smaller filter components are used in the boost converter to miniaturize the overall WEH system.

To experimentally verify the concept of the resistor emulation approach to perform MPPT for small-scale WEH, the input resistance of the boost converter, which is known as the emulated resistance of the wind turbine R_{em}, is

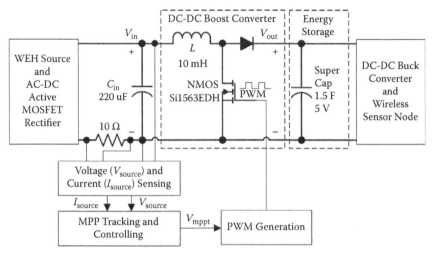

FIGURE 2.13
Overview of a DC-DC boost converter with MPPT.

electronically controlled to sweep through a wide range of values (10 to 800 Ω) as shown in the power curve (left side) and I-V curve (right side) of Figure 2.14. Based on the fundamental equations of the DC-DC boost converter under continuous conduction mode (CCM), the emulated resistance at the input port of the converter is governed by Equation 2.5:

$$R_{em} = (1 - D)^2 R_{load} \qquad (2.5)$$

Under a static load condition, a fixed resistor R_{load} of 1.2 kΩ is selected to represent the average power consumption of the wireless sensor load. Since R_{load} is a constant, the input resistance R_{em} of the converter is related to the duty cycle D of the boost converter as expressed in Equation 2.5. As such, R_{em} is manipulated to match with the internal resistance of the wind turbine of 150 Ω to attain MPPT by controlling the duty cycle D of the boost converter's gating signal.

$$D = 1 - \sqrt{\frac{R_{em}}{R_{load}}} \qquad (2.6)$$

Referring to the sweeping process captured in the oscilloscope as shown by the power curves in Figure 2.14, it is observed that all the maximum power points (MPPs) are centralized around the optimal resistance of 150 Ω for different wind speeds. The experimental results obtained illustrate that using the designed boost converter circuitry with a resistor emulation MPPT approach, maximum power is indeed transferred from the wind turbine generator to the wireless sensor load.

The MPPT performance of the designed boost converter together with its proposed closed-loop resistance emulator was also tested experimentally under dynamic conditions, that is, changing wind speeds from 2.3 to 6.3 m/s as

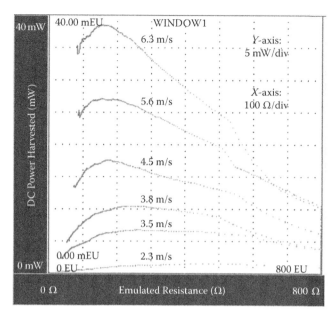

FIGURE 2.14
Experimentally obtained power and I-V curves for various incoming wind speeds.

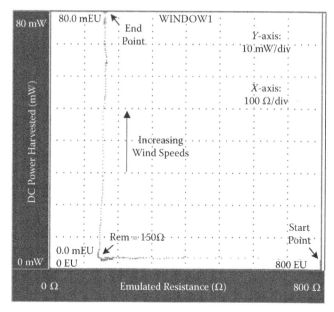

FIGURE 2.15
Performance of an MPP tracking boost converter under varying wind speeds.

illustrated in Figure 2.15. First, a light wind of 2.3 m/s was blown at the wind turbine generator, and the start point marked in Figure 2.15 was the initial condition of the WEH system. The MPP tracker utilizes the closed-loop PI controller to manipulate the duty cycle of the boost converter according to Equation 2.6, which in turn controls the input resistance of the boost converter towards (left side) the optimal resistance value of 150 Ω. Once the MPP of the power curve for a wind speed of 2.3 m/s is reached, the closed-loop resistance emulator controls the boost converter to maintain power harvested from the wind turbine generator for all the other MPPs occurring at different wind speeds. Referring to Figure 2.15, it is observed that the emulated resistance R_{em} is maintained at around 150 Ω with increasing wind speeds, until the MPP of the power curve for a wind speed of 6.3 m/s, marked as the end point in Figure 2.15, is reached. This shows that the dynamic condition of the environment is well taken care of by the proposed closed-loop resistor emulator of the designed boost converter.

The designed boost converter with the resistor emulation MPPT approach has already been demonstrated to yield excellent performance in extracting maximum power from the wind turbine generator, but this comes at the expense of additional power losses in the converter and its associated control, sensing, and PWM generation circuits. Thus, it is necessary to investigate the significance of these power losses as compared to the total harvested power. The first investigation is to determine the efficiency of the boost converter η_{conv} as a function of its output load power P_{load} over its input DC power P_{dc}. Taking the target deployment area with average wind speed of 3.62 m/s as an example, the efficiency of the converter is calculated to be as follows:

$$\eta_{conv} = \frac{P_{out}}{P_{in}} * 100\% = \frac{V_{out}^2 / R_{load}}{V_{in} I_{in}} * 100\%$$

$$= \frac{9.71 V^2 / 1200\Omega}{1.15V * 8.14 \text{ mA}} * 100\% = 84\% \tag{2.7}$$

For all other wind speeds, the efficiencies of the boost converter are calculated using Equation 2.7 to be between 80% and 90%, and the computed results are shown in Figure 2.16. As seen, even for a light wind speed condition where the power harvested is small (around 2 mW), the boost converter is still able to achieve a reasonably good efficiency of 86%. This exhibits the ability of the DC-DC boost converter to attain high efficiency in a very low power rating condition.

Another investigation being carried out is to determine the power consumption of the associated control, sensing, and PWM generation electronic circuits and its significance as compared to the harvested power. Based on the voltage and current requirements of each individual component in the sensing and processing circuits, the total power consumption of the electronic circuits is calculated to be

$$P_{consumed} = P_{sensing} + P_{processing} + P_{PWMgeneration} \tag{2.8}$$

$$= 3V * (74 \ \mu A + 15 \ \mu A + 30 \ \mu A) = 0.357 \text{ mW}$$

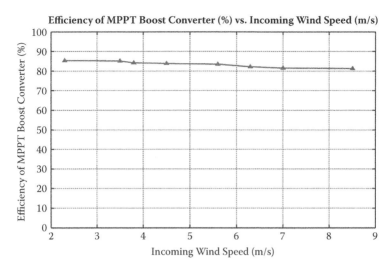

FIGURE 2.16
Efficiency of an MPPT boost converter for various incoming wind speeds.

Taking into account both the power loss across the boost converter and the power loss in the associated control, sensing, and PWM generation circuits mentioned in Equations 2.7 and 2.8, respectively, the performance comparison between the WEH system with MPPT and without MPPT are tabulated in the bar chart shown in Figure 2.17. For all the wind speed measurement points shown in Figure 2.17, it is observed that the performance of the WEH system with MPPT, including the converter's efficiency loss and circuits' power loss,

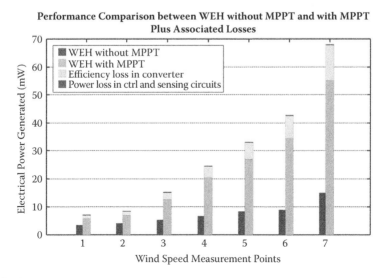

FIGURE 2.17
Performance comparison between the WEH system without MPPT and the WEH system with MPPT plus its associated losses for various incoming wind speeds.

is more superior than the WEH system without MPPT. It is even more obvious for higher wind speeds, as seen in Figure 2.17, where the difference in the harvested power between the WEH system with an MPPT scheme and without an MPPT scheme is significant; up to four times more electrical power can be harvested from the wind turbine at a wind speed of 8.5 m/s. This exhibits the importance as well as the contribution of implementing MPPT in the WEH system.

2.1.2.3 Energy Storage

For long-term deployment of the wireless sensor node, it is required to have an energy storage device such as a supercapacitor and batteries onboard the sensor node to sustain its operation throughout the lifetime. It is also crucial to ensure that this energy storage device has an operational life span of equivalent length or even longer so that the WSN lifetime is prolonged. Comparing between the choice of using supercapacitors or batteries as the energy storage for the WEH system, the supercapacitor has been chosen. The reason is that the supercapacitor exhibits several superior characteristics over the batteries that are useful for the WEH system. These characteristics include numerous full-charge cycles (more than half a million charge cycles), long lifetime (10 to 20 years operational lifetime), and high power density (an order of magnitude higher continuous current than a battery) to provide high instantaneous power to the sensor node during burst mode operation such as radio transmission [34].

Unlike the discrete capacitors, which have very small capacitance values in the microfarad range and usually are used in supply rails for decoupling purposes, the supercapacitor has a very large capacitance value in the farad range suitable for energy storage purposes. When a large capacitor that is in the farad range is initially attached to the energy source, the component with minimum energy stored acts as a short circuit to the energy source, and the supply rail voltage drops to the capacitor voltage level. The same situation occurs when the large capacitor is attached to the wind turbine generator as well. Although the wind turbine still charges the supercapacitor under this condition, it does not do so efficiently. This is because the charging process is not executed at the MPP, which is the voltage and current combination that maximizes power output under a given wind speed condition. A supercapacitor that is charged in this manner reduces the WEH efficiency by a factor of two to four as illustrated in Figure 2.18. Hence, the dynamic response of the supercapacitor is important to consider in the design of the boost converter to ensure constant MPPT operation is achieved by having a closed-loop resistor emulator instead of the open-loop resistor emulation method suggested [70] where the load impedance is assumed to be constant.

For a time period of 500 s as shown in Figure 2.18, the supercapacitor is charged by the WEH system from its discharged stage. At V_{cap} (500 s), the 1.5-F supercapacitor is being charged to voltage levels of 2.14 V with the MPPT scheme and 0.66 V without the MPPT scheme. Comparing between the two schemes, it is obvious that the charging performed by the WEH system

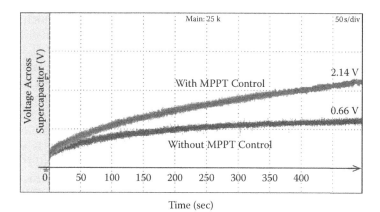

FIGURE 2.18
Performance of the WEH system with MPPT and without MPPT for charging a supercapacitor.

with MPPT is much higher than its counterpart without the MPPT scheme. This is because more electrical power is transferred from the wind turbine through the boost converter with the resistor emulation MPPT scheme to the supercapacitor. Under a dynamic load condition, changing R_{load}, the closed-loop resistance emulator is still able to manipulate the duty cycle D of the boost converter given by Equation 2.6 to maintain R_{em} always at the optimal resistance value, so that MPPT operation takes place. When the WEH system is operating with the MPPT scheme, the amount of energy accumulated in the supercapacitor after 500 s is 3.43 J, which is 10 times more than its counterpart of 0.33 J; hence, this exhibits the superior performance of the WEH system with the MPPT scheme over its counterpart under a dynamic load condition.

The charging process seen in Figure 2.19, is divided into two regions: uncontrollable and controllable. Initially, the supercapacitor is fully discharged,

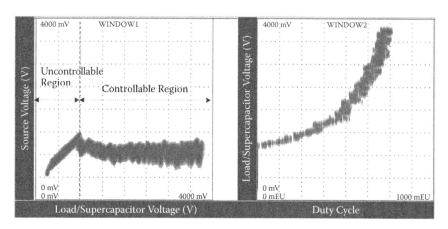

FIGURE 2.19
Illustration of the supercapacitor's charging process using the WEH system with MPPT.

so the WEH system charges the supercapacitor freely without much control over the duty cycle of the boost converter. The controllable region only starts when the supercapacitor voltage builds up to the MPPT voltage, $V_{mppt} = 1.15$ V, of the WEH system at a wind speed of 3.62 m/s. After this, the supercapacitor continues to be charged with the maximum power harvested from the WEH system. The whole process is illustrated by the source versus load voltage diagram (left side) of Figure 2.19. Based on the governing equation expressed in Equation 2.5, as the voltage of the supercapacitor charges, the effective resistance of the supercapacitor increases; hence, the duty cycle is adjusted according to Equation 2.6, as shown in Figure 2.19, to maintain MPPT at optimal resistance. By adding a supercapacitor, whose charging characteristics are nonlinear in nature, as an energy buffer between the source and the load, it is shown that the closed-loop resistance emulator of the boost converter is still able to operate the WEH system at its MPPs.

2.1.2.4 Wireless Sensor Nodes

The WEH system is designed to power a commercially available wireless sensor node supplied by Texas Instruments (TI) known as the wireless target board, eZ430-RF2500T. The operations of the wireless sensor node deployed in an application field are comprised of (1) sensing some external analog signals such as voltage and current signals of the wind turbine generator or temperature signal and (2) communicating and relaying the sensed information to the gateway node every 1 s. On receiving the data at the base station, the collected data is then postprocessed into usable information for any follow-up action.

Using the readily available wireless development tool eZ430-RF2500T for the TI wireless sensor node, which consists of an MSP430 microcontroller and a CC2500 RF transceiver, all the hardware and software required to develop the entire wireless project with the MSP430 is easily performed in a convenient USB (universal serial bus) stick. The eZ430-RF2500T uses the MSP430F2274 16-bit ultralow-power microcontroller, which has 32-kB flash, 1-K RAM, 10-bit analog-to-digital converter (ADC), and 2 op-amps and is paired with the CC2500 multichannel RF (radio-frequency) transceiver designed for low-power wireless applications. Since the MSP430 microcontroller is part of the wireless sensor node, it is very convenient to make use of the onboard microcontroller, without incurring much extra overhead power, to achieve the proposed MPPT scheme based on the resistor emulation approach rather than implementing the MPPT scheme with the analog circuit [36, 70].

Long-term operation is an important goal of the WSN system. One attempt to achieve this goal is to reduce energy consumption of the sensor node. Energy reduction is carried out by improving hardware design and more intelligent power management, which entails turning off unused components or slowing energy-hungry devices such as a microcontroller during idle periods. One approach taken is lowering the clock frequency from 1 MHz to 12 kHz, which uses the internal very-low-power, low-frequency oscillator (VLO) without requiring a crystal. The active mode supply current at 3-V supply voltage is tremendously reduced by a factor of 20, from 390 to 15 μA, respectively, thus reducing the power consumed. To compensate for the slow switching frequency, an external PWM generation circuit, which consumes few tens of microwatts, is designed and implemented to multiply the low-frequency PWM control signal generated from the low-power microcontroller to the 10-kHz range to miniaturize the size of the passive magnetic components. Another approach taken to reduce the energy consumption is duty cycling the transmission time of the energy-hungry radio module at a slower rate of every second preset for this chapter.

2.1.3 Experimental Results

The proposed concept of a self-powered WEH wireless sensor node using an efficient power management circuit, as illustrated in Figure 2.20, has been implemented in a hardware prototype for laboratory testing. Several tests are conducted during the experiments to validate the performance of the optimized WEH system using an AC-DC active rectifier and MPPT with a resistor emulation approach in sustaining the operation of the wireless sensor node.

2.1.3.1 Performance of WEH System with an MPPT Scheme

The experimental tests were conducted in accordance with the wind condition of the deployment ground illustrated in Figure 2.2, where the average wind speed is given as 3.62 m/s. Three experimental tests were conducted as shown by three different operating regions in Figure 2.21 to differentiate the performance of the WEH system and its resistor emulation MPPT scheme in powering the load consisting of a supercapacitor, sensing and control circuit, and wireless sensor node. The electrical load was first powered with a WEH system without MPPT, then with a WEH system with MPPT, and last without a WEH system and MPPT. Referring to Figure 2.21, it is observed that the supercapacitor voltage V_{cap} keeps decreasing during the period when the WEH system was not equipped with MPPT and even more obvious for the case where neither WEH system nor MPPT was integrated into the sensor node. This phenomenon indicates that being solely dependent on the energy storage or the electrical power harvested by the WEH system without an MPPT scheme is not sufficient to sustain the operation of the sensor node. It is only when the WEH system was incorporated with the MPPT scheme that sufficient power was provided for both the operation of the wireless sensor node and to charge the supercapacitor. Whenever the WEH system was operating in its MPPT mode (see Figure 2.21), the generated output voltage of the wind turbine V_{in} was controlled by the microcontroller to follow the MPPT voltage ($V_{mppt} = 1.15$ V) based on the resistor emulation algorithm.

The effect of MPPT on the WEH system was further examined using the waveforms as shown in Figure 2.22. First, it is observed that the voltage across the supercapacitor V_{cap} dropped from 2.9 to 2.75 V after 350 s. In a similar manner, the output DC voltage of the wind turbine V_{in} as shown in Figure 2.22

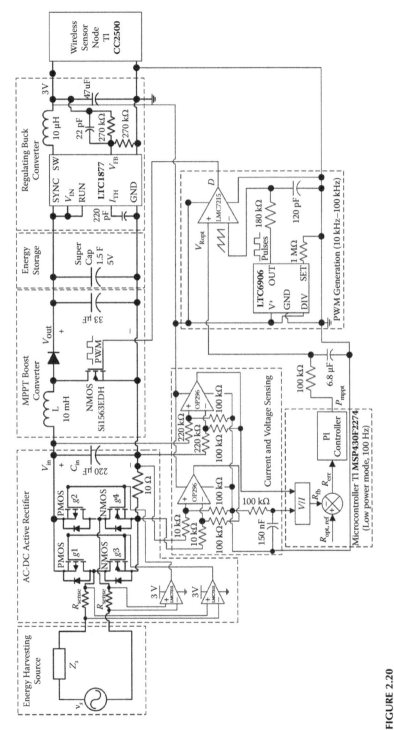

FIGURE 2.20
Schematic diagram of the self-powered wind sensor system.

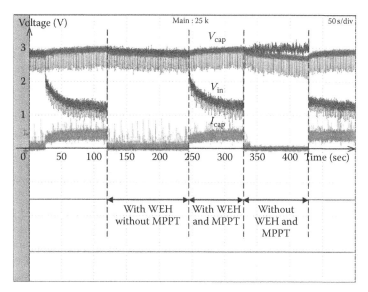

FIGURE 2.21
Operation of the sensor node under various powering schemes.

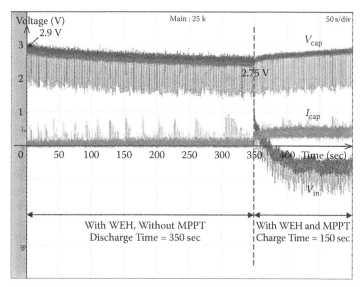

FIGURE 2.22
Performance of the WEH system with MPPT and without MPPT.

also dropped from 2.9 to 2.75 V. Based on the characteristic curve of the wind turbine generator shown in Figure 2.11, the amount of electrical power harvested by the WEH system at a wind speed of 3.62 m/s and V_{in} of between 2.75 and 2.9 V was around 1 to 2 mW. Since the power consumption of the wireless sensor node, including the sensing and control circuit (~0.4 mW), of around 3.6 mW at 1 s per transmission was more than the harvested power of 1 to 2 mW, the power harvested by the WEH system alone was not sufficient to maintain the operation of the sensor node. However, when the source and load impedances were nicely matched using the boost converter and its closed-loop resistance emulator, the harvested power from the wind turbine increased tremendously, generating sufficient power to charge the supercapacitor V_{cap} from 2.75 back to 2.9 V in 150 s, as shown in Figure 2.22, which is less than half of the previous discharge time. At a wind speed of 3.62 m/s, the electrical power harvested by the WEH system with its MPPT mode was approximately 7.86 mW, even if we consider the losses in the converter where 3.6 mW of power were consumed by the wireless sensor node, the rest of the harvested power of 4.26 mW was supplied to charge the supercapacitor. As such, the WEH system with the MPPT scheme was definitely able to sustain the sensor node's operation.

Another experimental test was carried out to compare the performance of a conventional sensor node, which operates solely on the energy storage, and a WEH sensor node as shown in Figure 2.23. As the conventional sensor node consumes 3.6 mW of average power from the supercapacitor, the voltage across the supercapacitor dropped from 2.8 to 2.55 V in around 275 s, which is calculated to be 1 J of energy transferred to the load. Compared to the WEH system without MPPT, as shown in Figure 2.22, the discharge rate of the supercapacitor was much higher in this case because there was no extra power generated from the WEH system to supplement the sensor node's operation. Once the WEH system with MPPT was activated, the harvested power of 7.86 mW was used to power the sensor node as well as to charge the supercapacitor back to 2.8 V in 225 s.

Among the three testing options, the WEH wireless sensor node with the MPPT scheme yielded the most superior performance. This is because the WEH sensor node was incorporated with a wind energy supply designed to harvest power at its optimal point, to charge the supercapacitor and sustain the operation of the sensor node. To further validate the superior performance of the WEH wireless sensor node with the MPPT scheme incorporated, the sensor node was tested in a light wind condition of 3-m/s wind speed where 5 mW of power was available at the output of the active rectifier as observed in Figure 2.9. After taking the losses of the DC-DC converter into consideration, the available power of 3.74 mW had a surplus after taking into account the power consumption of the wireless sensor node. Hence, the charging voltage V_{cap} waveform as shown in Figure 2.24 illustrates that the WEH system with the MPPT scheme, operating at lower wind speeds, was still able to supply power to the sensor node as well as to charge the supercapacitor.

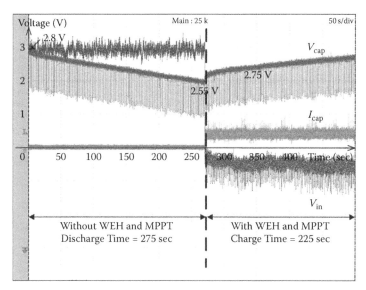

FIGURE 2.23
Performance comparison between a conventional sensor node and a WEH sensor node.

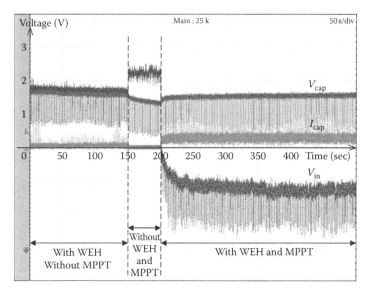

FIGURE 2.24
Operation of a sensor node at a light wind speed of 2.3 m/s.

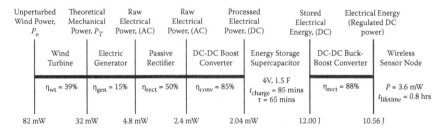

FIGURE 2.25
Line diagram of the power distributed in the WEH system without an active rectifier and MPPT scheme.

2.1.3.2 Power Conversion Efficiency of the WEH System

The WEH system as a whole is a complex system that is made up of many different subsystems. In order to understand how to improve the overall efficiency of the WEH system, it is important to study the performance of each of these subsystems and perform some power analysis to better understand how power is distributed at each power conversion stage of the WEH system. Line diagrams are drawn in Figures 2.25 and 2.26 to illustrate the input and output power available for each subsystem so that the power conversion efficiency of the subsystem can be determined.

Referring to Figure 2.25, the line diagram starts from the input with a wind speed of 3.62 m/s where 82 mW of raw wind power is supplied to the wind turbine with an efficiency of 39%, and 32 mW of mechanical power is available for harvesting. Using a standard diode-based full-bridge rectifier, 50% of the raw electrical power (AC) generated at the output of the wind turbine generator of approximately 4.8 mW is converted into raw electrical power (DC) of 2.4 mW. With a standard boost DC-DC converter being used in the WEH system, it is observed in Figure 2.25 that the electric generator has a relatively low efficiency (15%) because its internal source impedance is not properly matched with the subsequent subsystems. Hence, the processed DC power to charge the supercapacitor is only 2.04 mW.

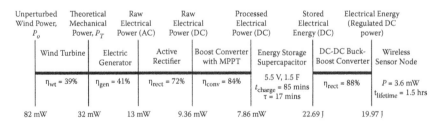

FIGURE 2.26
Line diagram of the power distributed in the proposed WEH system.

Another line diagram, shown in Figure 2.26, is drawn to illustrate the power distribution in the WEH system with an active rectifier and a resistance emulation MPPT scheme. When an active rectifier is employed, the AC-DC conversion efficiency has been improved from 50% to 72%. The WEH system is further enhanced by using the closed-loop resistance emulator to perform impedance matching, and it is shown in Figure 2.26 that there is about three times more raw electrical power (AC) harvested from the wind turbine and then converted into electrical power (DC) of 7.86 mW to charge the supercapacitor. The performance comparison between the line diagrams exhibited in Figures 2.25 and 2.26 illustrates the significant contribution of the proposed power management circuit in the overall WEH system incorporated into the wireless sensor node.

For a 1.5-F, 5.5-V supercapacitor, the maximum amount of energy stored in the supercapacitor is 22.69 J at 5.5 V. To fully charge the supercapacitor, the required charging time t_{charge} is computed to be 85 min when 7.86 mW of electrical power is supplied. To make a fair comparison, the same charging time of 85 min is used, and the WEH system with a standard power management circuit is able to transfer 12 J of energy into the supercapacitor, which is about half the maximum capacity of the supercapacitor. Comparing the two line diagrams shown in Figures 2.25 and 2.26, the operational lifetime of the sensor node powered by the WEH system with an MPPT scheme is twice that of the sensor node powered by the WEH system with a standard power management circuit, hence making the WEH system with the MPPT scheme a viable solution for extending the lifetime of the WSN.

2.1.4 Summary

There is a need for a paradigm shift from the battery-operated conventional wireless sensor node towards a truly self-autonomous and sustainable energy harvesting wireless sensor node. Small-scale WEH through a micro wind turbine generator is one of the options to power small autonomous sensors deployed in remote locations for sensing under long-term exposure to a hostile environment such as a forest fire. Two challenging problems associated with a small-scale WEH, such as rectification of low-amplitude AC voltage and impedance mismatch between source and load, have been addressed. An efficient power management unit of the WEH system, consuming very little power (0.447 mW), has been designed to overcome these challenges. It has been demonstrated with experimental results that with the proposed active rectifier and resistance emulation-based MPPT scheme, more electrical power (from 2.04 to 7.86 mW) is harvested from the wind turbine with higher overall power conversion efficiency from 2.5% to 9.6%. As such, it is more viable to achieve a truly self-autonomous and sustainable wireless sensor node with optimal WEH using an efficient power management circuit.

2.2 Indirect WEH Approach Using Piezoelectric Material

With the great advancement of microelectronic technologies in many areas such as integrated circuit (IC) designs, chip fabrications, and so on, the power requirement for wireless sensor nodes continues to decrease from the milliwatt to microwatt level. This paves the way for micro-WEH for some WSN applications, whereby there is a need for the small-scale WEH system to be as small as possible and highly portable so it does not interfere with the normal operation of the deployment area. The direct WEH approach using a wind turbine generator, as illustrated in Section 2.1, has been successfully demonstrated in powering the miniaturized wireless sensor node. However, there are certain limitations with this type of conventional wind power generator, which uses a large rotational turbine of 3-cm blade radius to harvest energy from the wind flow. Limitations include: a large wind front contact with the wind is required for good WEH, mechanical constraint on the miniaturization of the electric generator and gearbox (if any), and more.

The physical size of wind turbine generators that have been reported in Section 2.1 as well as the academic literature is still relatively bulky compared to the sensor node. Weimer et al. present a compact anemometer-based solution [60] for remote area WEH; the space needed by the anemometer is still relatively large compared to the miniature sensors. Another research study by Priya et al. [61] describes the "piezoelectric windmill" that consists of several piezoelectric actuators arranged along the circumference of the mill in the cantilever form. The design of this windmill is very large and complex such that the power generated, which is in the milliwatt range, exceeds the power requirement of the ultralow-power sensor node. Hence, these large and bulky WEH systems are not cost effective and appropriate for WSN applications, which require miniaturized devices.

In this section, an indirect WEH approach using piezoelectric material is proposed for powering a miniaturized wireless sensor node. The proposed piezoelectric wind energy harvester is very different from the conventional bladed wind turbine; a novel way of harvesting wind energy through the piezoelectric effect is explored to address the limitations of the wind turbine generator.

2.2.1 Vibration-Based Piezoelectric Wind Energy Harvester

A novel way of harvesting electrical energy from wind energy using bimorph piezoelectric material is proposed here. The novel piezoelectric wind energy harvester collects the vibration energy induced from the wind flow, and the vibration creates stress on the piezoelectric material to generate electrical energy. The application of this novel piezoelectric-based wind energy harvester is similar to the wind turbine research work, which is for a low-powered autonomous wind speed sensor. The piezoelectric wind harvester is a unique system that combines the concept of wind and piezoelectric effect. Making use of the force generated by the flow of wind to vibrate the piezoelectric material, the mechanical energy harvested from the vibration of the piezoelectric material is converted into electrical energy. The advantages of the piezoelectric wind energy harvester are that it is compact and sensitive to low-speed wind. Although the amount of energy that can be harvested is quite limited, it is shown experimentally in the section that the piezoelectric energy harvester is sufficient to power the wireless electronic circuits to transmit five digital 12-bit signals to the remote base station. Once the trigger signal is received, the warning siren would be activated.

Figure 2.27 illustrates the power conversion process of the proposed vibration-based piezoelectric wind energy harvester, starting from the incoming wind flow v to the output electrical power generation P_{elec}. The power conversion process can be divided into three main stages: aerodynamic stage (Section 2.2.1.1), cantilever bending beam stage (Section 2.2.1.2), and piezoelectric stage (Section 2.2.1.3). Referring Figure 2.27, it can be seen that each of these power conversion stages has its own set of representative analytical models, which are elaborated in detail in subsequent sections. Before that, the operating principle of the vibration-based piezoelectric wind energy harvester, as illustrated in Figure 2.27, is first discussed and explained:

- Difference in wind speeds above v_a and below v_b the airfoil/blade creates a net pressure P that results in a lift force F at the tip of the blade. This phenomenon can be explained based on Bernoulli's principle.
- Net pressure P is applied at the tip ($x = L$) of the piezoelectric wind energy harvester, which is free to vibrate, and its other end is fixed with a clamp ($x = 0$). Due to the wind v flowing across the piezoelectric wind energy harvester, the tip of the harvester deflects up and down (δ_L); hence, vibration is generated on the piezoelectric material. This phenomenon can be explained with the Euler-Bernoulli cantilever beam theory.
- As the blade of the harvester swings, the piezoelectric material, which is bonded to the blade, experiences the mechanical stress σ of the generated vibration. Electrical AC power is thus harvested from the vibration-based piezoelectric wind energy harvester. This phenomenon can be explained using piezoelectricity theory.

2.2.1.1 Aerodynamic Theory

The aerodynamic effect of airflow on the piezoelectric wind energy harvester is described based on a well-known fluid mechanic principle: Bernoulli's principle. Bernoulli's principle states that in fluid flow, an increase in velocity occurs simultaneously with a decrease in pressure. This principle is a simplification of Bernoulli's equation, which states that the sum of all forms of energy in a fluid flowing along an enclosed path is the same at any two points in that

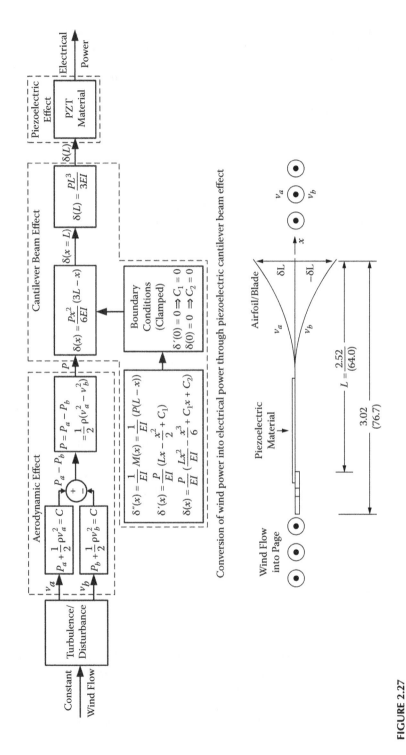

FIGURE 2.27

A diagram of the vibration-based piezoelectric wind energy harvester.

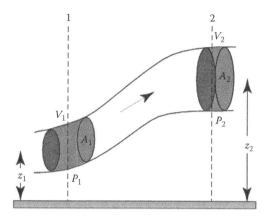

FIGURE 2.28
A diagram of a pipe through which an ideal fluid is flowing at a steady rate.

path. The fluid can be either a liquid or a gas, but for Bernoulli's principle to be applicable, the fluid is assumed to have the following qualities [75]:

- Fluid flows smoothly.
- Fluid flows without any swirls (also known as eddies).
- Fluid flows everywhere throughout the pipe (which means there is no "flow separation").
- Fluid has the same density everywhere (it is "incompressible" like water).

To understand how and why Bernoulli's principle works, the development of the relationship of the static and dynamic pressures using Bernoulli's equation has to be investigated. Bernoulli's equation along a streamline can be summarized as follows:

$$P_1 + \frac{1}{2}\rho v_1^2 + \rho g z_1 = P_2 + \frac{1}{2}\rho v_2^2 + \rho g z_2 = a\ constant \qquad (2.9)$$

where:
1 is the first point along the pipe in Figure 2.28
2 is the second point along the pipe in Figure 2.28
P is the static pressure of the fluid (Pa)
ρ is the density of the fluid (kg/m^3)
v is the velocity of the fluid (m/s)
g is gravitational acceleration (m/s^2)
z is height (m)

Referring to Bernoulli's principle applied for the airfoil case shown in Figure 2.29, the different points that fall along the same streamline flow of the wind (i.e., the effects due to gravity) are small compared to the effects due to kinematics and pressure, $z_1 \approx z_2$; hence, the $\rho g z$ term in Equation 2.9

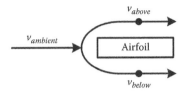

FIGURE 2.29
Different points along the same streamline for the application of Bernoulli's equation.

can be cancelled out on both sides. Referring to Figure 2.28, the point labelled as 1 is a point on the streamline far in front of the airfoil labelled in Figure 2.29 as ambient wind. The static pressure at point 1 is thus equated as $P_1 = P_{ambient}$. Similarly, by representing point 2 labelled in Figure 2.28 to be at a point above the surface of the airfoil as observed in Figure 2.29, Equation 2.9 can be expressed as

$$P_{ambient} + \frac{1}{2}\rho v_{ambient}^2 = P_{above} + \frac{1}{2}\rho v_{above}^2 = a \ constant \qquad (2.10)$$

In another condition below the surface of the airfoil, point 1 seen in Figure 2.28 is again a point on the streamline in front of the airfoil in Figure 2.29, and the values of $P_{ambient}$ and $v_{ambient}$ and the wind parameters are the same as the above-airfoil condition. Point 2 labelled in Figure 2.28 is represented by a point below the surface of the airfoil observed in Figure 2.29. Hence, Equation 2.10 is restructured into

$$P_{below} + \frac{1}{2}\rho v_{below}^2 = P_{above} + \frac{1}{2}\rho v_{above}^2 = a \ constant \qquad (2.11)$$

For horizontal wind flow, an increase in the wind speed would result in a decrease in the static pressure. As such, when the air flowing over the top of the airfoil v_{above} travels faster than the air flowing under the airfoil v_{below}, there is less pressure on the top P_{above} than on the bottom P_{below}, resulting in the net pressure expressed as

$$P = P_{above} - P_{below} = \frac{1}{2}\rho\left(v_{below}^2 - v_{above}^2\right) \qquad (2.12)$$

This resultant net pressure P is the input variable of the next power conversion stage of the piezoelectric wind energy harvester, which is the cantilever bending beam stage, described in the next section.

2.2.1.2 Cantilever Beam Theory

The main objective is to determine how much the vibration-based piezoelectric wind energy harvester would vibrate when wind flows past the harvester. As the positive/negative net pressure P generated due to the aerodynamic effect of the blade swing is loaded on the tip of the cantilever beam, the beam

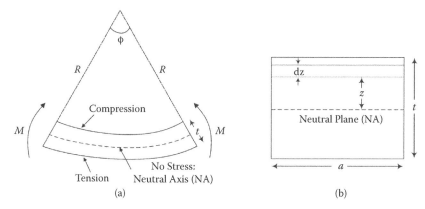

FIGURE 2.30
Section of the beam subjected to pure bending.

deflects downward/upward, respectively. The relationship between the net pressure created by the wind flow v and the beam deflection y is investigated and discussed. Once this relationship is established and the wind speed flowing above and below the wind energy harvester is known, the amount of vibration being generated on the piezoelectric material can then be calculated.

In this research, the vibration of the piezoelectric wind energy harvester is examined based on a simple beam theory studied in structural mechanics to understand the beam deflection behaviour. When a beam is subjected to a bending moment M of small deflections as shown in Figure 2.30, the outside of the bend is stretched, the inside is compressed, and in between them, there is the neutral axis or neutral plane, which does not experience any tensile stress at all [76]. The bending process of a section of the beam exhibited in Figure 2.30 is subjected to pure bending. For a homogeneous and symmetrical material, the neutral axis should be located at the geometrical centre.

Considering an element of the beam as shown in Figure 2.30a for bending analysis, the radius of curvature of the neutral axis NA is R, and the element of the beam includes an angle ϕ at the centre of curvature. An incremental change in the distance z from the NA has the length of $(R \pm z)\,\phi$ along the NA, so the extension/compression of the incremental distance is $\pm z\phi$, and the strain is $\pm z\phi/R\phi = \pm z/R$. Now consider a symmetrical beam with thickness t and width w pointing out of the page as shown in Figure 2.30b. When the symmetrical beam bends, the rectangle shown in Figure 2.30b would be rotated, and the bending moment needed can be calculated by considering the tensile forces involved. Since the strain ϵ and stress σ on the element dz caused by the bending are $\pm z/R$ and $E_y(z)^*\epsilon$, respectively, the force F required to achieve the moment M can be expressed as

$$F = (stress)(area) = E_y(z)\left(\frac{\pm z}{R}\right) a\; dz \qquad (2.13)$$

where $E_y(z)$ is the Young's modulus of the beam as a function of its thickness z. With reference to Figure 2.30b, the moment of the force about the neutral plane NA is given as

$$F.z = \frac{\pm E_y(z)z^2 a \, dz}{R} \tag{2.14}$$

The total bending moment on the rectangle beam is the sum of the individual moments on all the elements dz [76], which can be expressed as

$$Bending\ Moment = \frac{a}{R} \int_{-t/2}^{t/2} E_y(z)z^2 dz \tag{2.15}$$

This bending moment is equal to the applied moment at each point in the beam (which depends on the applied force and its distance from the specific point). The geometric moment of inertia I, which describes the effect of the rectangular shape of the beam cross section, is given as [77]

$$I = \int a z^2 dz \tag{2.16}$$

$$I = a \left[\frac{z^3}{3} \right]_{\frac{-t}{2}}^{\frac{t}{2}} = \frac{at^3}{12} \tag{2.17}$$

Now, consider the cantilever beam as shown in Figure 2.31; the beam has a length of L, width of w, and thickness of t with one of the ends fixed in cantilever form. The cantilever beam is homogeneous and symmetrical and has a constant value of E_y. A force F is applied to the free end of the beam, and the amount of deflection y at a distance L from the fixed end can then be determined. The radius of curvature R is defined as [76]

$$\frac{1}{R} = \frac{\frac{d^2 y}{dx^2}}{\left(1 + \left(\frac{dy}{dx}\right)^2\right)^{\frac{3}{2}}} \tag{2.18}$$

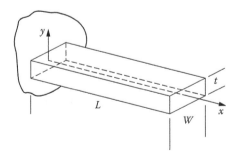

FIGURE 2.31
A cantilever beam.

For small deflections, $\frac{dy}{dx}$ is small; hence $(\frac{dy}{dx})^2 << 1$, and therefore the radius of curvature can be simplified to

$$\frac{1}{R} = \frac{d^2y}{dx^2} \tag{2.19}$$

Next, the bending moment at some distances x from the wall is equated to the applied bending moment shown in Equation 2.15. Using Equations 2.15, 2.16, and 2.19, the bending moment derived in Equation 2.15 can be described in a simpler form as follows:

$$Bending\ Moment = E_y I \frac{d^2y}{dx2} = F(L - x) \tag{2.20}$$

From the simplified bending moment equation, the beam curvature can be found as

$$\frac{d^2y}{dx2} = \frac{F}{E_y I}(L - x) \tag{2.21}$$

Integrate Equation 2.21 with respect to x once to derive the beam slope, which is expressed as

$$\frac{dy}{dx} = \frac{F}{E_y I}\left(Lx - \frac{x^2}{2}\right) + K \tag{2.22}$$

Based on the boundary condition, which states that at $x = 0$, $dy/dx = 0$, so $K = 0$. Integrating the beam curvature equation a second time gives the beam deflection

$$y = \frac{F}{E_y I}\left(\frac{Lx^2}{2} - \frac{x^3}{6}\right) + K' \tag{2.23}$$

There is a boundary condition that, at $x = 0$, $y = 0$, so $K' = 0$. Hence, at $x = L$, the tip deflection is found to be

$$y_L = \frac{FL^3}{3E_y I} \tag{2.24}$$

The relationship between the aerodynamic force F created by the wind flow and the tip deflection of the piezoelectric wind energy harvester is established. When a positive/negative net force is loaded on the tip of the piezoelectric wind energy harvester mounted in cantilever form, the tip downward/upward deflection of the harvester can be estimated using Equation 2.24. A summary of the theoretical and experimental deflections of the piezoelectric wind energy harvester with respect to various wind speeds is given in Table 2.1.

Figure 2.32 exhibits the experimental setup used to measure the tip deflection of the piezoelectric wind energy harvester when wind flows across the

TABLE 2.1

Relationship between Incoming Wind Speed and Tip Deflection of the Cantilever Piezoelectric Wind Harvester Beam

v_a (m/s)	v_b (m/s)	Force (N)	I (m⁴)	Young's Modulus of Mylar/Polyethylene Terephthalate, E_y (GPa)	Theoretical Deflection (mm)	Experimental Deflection (mm)
7	1	0.0134	1.29E-14	2.5	6.49	6.5
5	2	0.0059	1.29E-14	2.5	2.84	3.0
3	1	0.0022	1.29E-14	2.5	1.08	1.0

harvester. It can be seen from Figure 2.32 that there is a measurement ruler held vertically by a square clamp beside the piezoelectric wind energy harvester. The ruler acts like a measurement scale to determine the magnitude of the deflection at the tip of the harvester, and the neutral/starting point of the measurement scale has been offset by 9 cm. A silver electric fan is placed in the background of the four figures to simulate a wind source to the wind piezoelectric energy harvester. The wind flow generated by the electric fan excites the harvester to vibrate, and the amount of deflection observed at the tip of the harvester is read from the measurement ruler. Figures 2.33, 2.34, and 2.35 illustrate the three experiments carried out for incoming wind speeds of $V_a = 7$ m/s and $V_b = 1$ m/s, $V_a = 5$ m/s and $V_b = 2$ m/s, and $V_a = 3$ m/s and $V_b = 1$ m/s, respectively. The experimental deflections read from the three figures are 6.5 mm, 3 mm, and 1 mm, respectively, and they are recorded in Table 2.1 to verify the theoretical value of the beam deflection calculated using Equation 2.24.

FIGURE 2.32
Piezoelectric wind energy harvester under no wind speed.

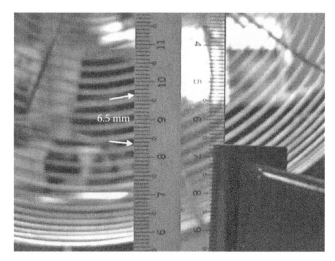

FIGURE 2.33
Piezoelectric wind energy harvester under a wind speed of $V_a = 7$ m/s and $V_b = 1$ m/s.

Referring to Table 2.1, the percentage of deviation between the calculated theoretical and measured experimental values for the tip deflections of the piezoelectric wind energy harvester is found to be less than 1%. This shows that the proposed relationship developed between Bernoulli's aerodynamic theory and the cantilever beam theory is pretty accurate and established. In addition, the derived equations, starting from the incoming wind speed differences to the beam deflection, are good representations of the experimental

FIGURE 2.34
Piezoelectric wind energy harvester under a wind speed of $V_a = 5$ m/s and $V_b = 2$ m/s.

FIGURE 2.35
Piezoelectric wind energy harvester under a wind speed of $V_a = 3$ m/s and $V_b = 1$ m/s.

results. Hence, for further expansion in the research work, these derived equations can be used as the baseline to predict and estimate the performance of the piezoelectric wind energy harvester for the required specification and operating condition. Furthermore, with good understanding on the working concept and technical derivation of the piezoelectric wind energy harvester, the present design of the harvester can be easily altered and optimized for different operating conditions. Take, for instance, by altering the geometric moment of inertia I discussed in Equation 2.16, which describes the effect of the rectangular shape of the beam cross section, the tip deflection of the harvester is expected to change. The remaining step is to find the relationship between the deflection of the cantilever wind energy harvester and the electric power generation from the piezoelectric material.

2.2.1.3 Piezoelectric Theory

For a piezoelectric generator mounted in cantilever form, the transverse mode (mode 31) of operation is considered. The mechanical force F applied on the piezoelectric generator is perpendicular to its output electrodes; hence, the surface where the charge is collected and the surface where the force is applied are independent. According to Smits, Dalke, and Cooney [78] and Wang et al. [80], for a series-connected bimorph bender subjected to the excitations of an electric voltage V across its thickness, a uniformly distributed external body load p, an external tip force F perpendicular to the beam, and an external moment M at the free end, the generated electrical charge can be expressed as

$$Q = \frac{3d_{31}L}{t^2}M + \frac{3d_{31}L^2}{2t^2}F + \frac{d_{31}wL^3}{2t^2}p + \frac{\epsilon_{33}^X Lw\left(1 - k_{31}^2/4\right)}{t}V \qquad (2.25)$$

where L, w, and t are the bimorph piezoelectric generator length, width, and thickness, respectively. ϵ_{33}^X is the dielectric constant of the piezoelectric material under a free condition; d_{31} is the transverse piezoelectric coefficient, and k_{31} is the transverse piezoelectric coupling coefficient. When only an external tip force F is acting on the tip of the bimorph piezoelectric generator ($x = L$), the generated electric charge in the bimorph defined as Equation 2.25 by Wang et al. [80], becomes

$$Q = \frac{3d_{31}L^2}{2t^2}F \qquad (2.26)$$

For a bimorph piezoelectric generator, two pieces of piezoelectric material are mechanically bonded together as a whole; hence, the overall dielectric constant of the bimorph piezoelectric generator is smaller than the free dielectric constant of only the piezoelectric material. As such, the overall dielectric constant ϵ_b is given as

$$\epsilon_b = \epsilon_{33}^X(1 - k_{31}^2/4) \qquad (2.27)$$

The total capacitance of the bimorph piezoelectric generator can be obtained by

$$C = \frac{(overall\ dielectric\ constant)(electrode\ surface\ area)}{thickness\ separating\ the\ electrodes} \qquad (2.28)$$

$$C = \frac{\epsilon_{33}^X Lw(1 - k_{31}^2/4)}{t} \qquad (2.29)$$

Based on Equations 2.26 and 2.28, the open circuit electric voltage V_{oc} generated by the bimorph piezoelectric generator when an external tip force is applied is expressed as

$$V_{oc} = \frac{Q}{C} = \frac{3d_{31}L}{2\epsilon_{33}^X wt(1 - k_{31}^2/4)}F \qquad (2.30)$$

The dielectric constant ϵ_{33}^X, piezoelectric constant d_{31}, and coupling coefficient k_{31}^2 of the piezoelectric material and the dimensions of the bimorph are given by the vendor in the technical datasheet. The properties of the piezoelectric material are given in Table 2.2.

Alternatively, the open circuit electric voltage V_{oc} generated by the bimorph piezoelectric generator can also be a function of the tip deflection y_L. Rearranging Equation 2.24, it can be seen that the external tip force F is related to the tip deflection of the piezoelectric generator by

$$F = \frac{3E_y I y_L}{L^3} \qquad (2.31)$$

Since the Young modulus E_y of the piezoelectric material is known in Table 2.2 and the moment of inertia I of the cantilever piezoelectric generator can be calculated using Equation 2.16, the open circuit electric voltage V_{oc}

TABLE 2.2

Properties of Piezoelectric Material

Description of Properties	Unit	Symbol	Value
Piezoelectric material	Piezo systems	PSI-5A4E	Lead zirconate titanate
Piezoelectric strain/field coefficient	Metres/volt	d_{33}	390×10^{-12}
Piezoelectric charge density/stress coefficient	Coulombs/newton	d_{31}	-190×10^{-12}
Piezoelectric strain/charge density coefficient	Metres/coulomb metre	g_{33}	24×10^{-3}
Piezoelectric field/stress coefficient	Volt/newton metre	g_{31}	-11.6×10^{-3}
Coupling coefficient		k_{31}	0.35
Elastic modulus	Newtons/metre^2	Y_{E1}	6.6×10^{10}

generated by the vibration-based piezoelectric wind energy harvester can be estimated based on the tip deflection y_L of the piezoelectric generator, which is expressed as

$$V_{oc} = \frac{9d_{31}E_y I}{2\epsilon_{33}^X L^3 wt\left(1 - k_{31}^2/4\right)} y_L \qquad (2.32)$$

2.2.2 Characteristics and Performances of a Piezoelectric Wind Energy Harvester

Wind flowing with a certain range of speeds is able to stir a certain amount of vibration on the plastic flapper attached to the piezoelectric material, which thus passes the vibrational energy to the piezoelectric material. The harvested vibrational energy from the piezoelectric material would then be converted into electrical energy at the output of the piezoelectric wind energy harvester.

The relationships between the harvested electrical energy and the incoming wind depends on the orientations and rotating angles between the piezoelectric wind energy harvester with respect to the directions of the incoming wind flow as shown Figures 2.36 to 2.39. Taking the concept about the flight dynamics of a vehicle rotating in three dimensions around its coordinate system origin, the three rotating angles defined in this research work (θ_L, θ_W, and θ_Z) are equivalent to roll, pitch, and yaw angles, respectively. Experiments have been conducted to find out the amount of electrical power that is generated by the piezoelectric wind energy harvester when placed in different orientations.

Figures 2.37, 2.38, and 2.39 illustrate the power curves of the piezoelectric wind energy harvester with different rotating angles θ_L, θ_W, and θ_Z ranging from 0° to 90°. Similar sets of experimental tests have been collected for the rotating angles ranging from −90° to 0°, and the results are quite symmetrical.

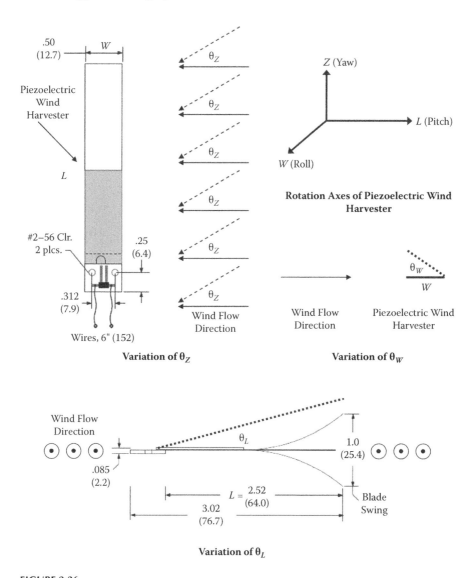

FIGURE 2.36
Orientations and rotating angles of the vibration-based piezoelectric wind energy harvester.

Therefore, experimental results are provided only for rotating angles ranging from 0° to 90°. From the three figures, one common phenomenon that can be observed: As the rotating angle increases from 0° to 90°, the amount of electrical power generated by the piezoelectric wind energy harvester decreases. This phenomenon is due to the reduction in the net lift force created by the pressure difference between the top and bottom surfaces of the

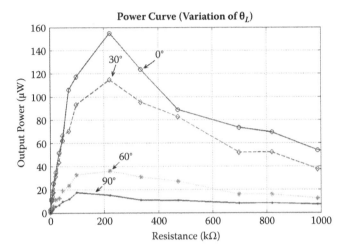

FIGURE 2.37
Power generated by the harvester with angles of θ_L.

piezoelectric wind energy harvester. Bernoulli's equation in fluid dynamics states that as the speed of a fluid (air/water) flow increases, its pressure decreases. Hence, it is clear that the pressure differences around the piezoelectric wind energy harvester are brought about by the variations in wind speed (caused by the disruption and turning of the air flowing past the piezoelectric wind energy harvester) at all points of the piezoelectric wind energy harvester. From these experimental tests, it can be concluded that to get the

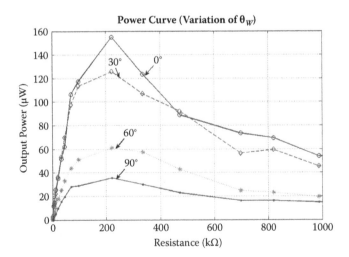

FIGURE 2.38
Power generated by the harvester with angles of θ_W.

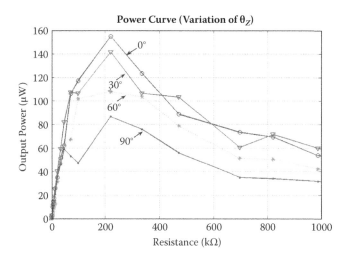

FIGURE 2.39
Power generated by the harvester with angles of θ_α.

maximum power output, θ_L, θ_W, and θ_Z should be set at $0°$ with respect to the wind flow direction.

Piezoelectric-based wind energy harvester offers several advantages over the conventional wind turbine harvester. These include instant starting with no dead time (due to the inertia of the wind turbine generator); small size and ultralight weight; extremely low magnetic permeability (suitable for use in high-magnetic-field environments); and almost no heat dissipation. Therefore, the piezoelectric wind energy harvester is suitable to power the miniaturized autonomous sensor used in many application areas, such as structural and automation applications. To study the feasibility of using the piezoelectric wind energy harvester as a generator to supply power to the power conditioning circuit and the RF transmitter for practical application, some characterizations have been conducted to better understand the capability of the piezoelectric wind energy harvester. One of the characterization works carried out on the harvester was to determine the relationship between the open circuit rectified DC voltage and the wind speed as shown in Figure 2.40.

Based on the plot in Figure 2.40, the DC output voltage of the piezoelectric wind energy harvester can be used to predict the speed of the incoming wind. For a given wind speed (any wind speed beyond this threshold value would trigger a 12-bit digital signal to be transmitted to the base station in a wireless manner), the corresponding preset voltage level for the power management circuit can be determined from Figure 2.40. Take, for instance, at a wind speed of 6.7 m/s, the DC output voltage of the piezoelectric wind energy harvester that is measured to be around 8.8 V. If this is the preset wind speed threshold for sounding the siren at the remote base station, then the

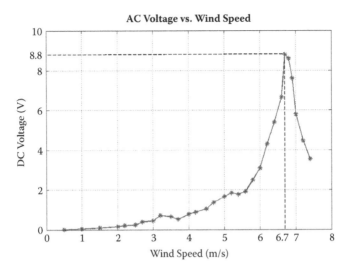

FIGURE 2.40
Open circuit AC voltage over a range of wind speeds.

DC output voltage of 8.8 V is used to power the wireless RF transmitter to transmit several 12-bit digital signals to the base station in a wireless manner. Another characterization illustrated by Figure 2.41 shows the typical output electrical power characteristic of the piezoelectric wind energy harvester as a function of the load resistances at different wind speeds. Simply, the figure shows how much power the piezoelectric wind energy harvester can supply continuously to the load.

Note that in Figure 2.41 a constant load resistance of 220 kΩ results in a maximum output power over the full range of wind speeds. At a wind speed of 6.7 m/s and the load resistance fixed at 220 kΩ, Figure 2.41 shows that the maximum amount of output power that can be harvested from one piezoelectric wind energy harvester is around 155 μW. If the number of piezoelectric wind energy harvesters is to be increased, then the amount of power that can be generated would be a multiple of 155 μW. With the present microelectronic technology, to power the sensor continuously with 155 μW is quite impossible; hence, an energy storage with some triggering circuit has to be included in the power management system to ensure that the electric energy stored in the storage element is sufficient before supplying the power to the RF wireless transmitter and its associated power management circuit.

2.2.3 Power Processing Units

The harvested power of the piezoelectric wind energy harvester is first fed into the power processing unit (PPU), and then the regulated output voltage from

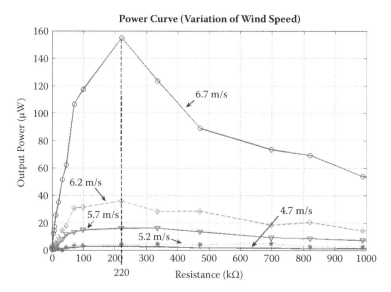

FIGURE 2.41
Piezoelectric wind energy harvester power source curves over a range of wind speeds.

the PPU is used to power the RF transmitter. The main function of the PPU is to convert and condition the unregulated raw electrical voltage into usable regulated voltage for the RF transmitter load. In the PPU, the unregulated raw power needs to undergo three different stages of power conditioning before the usable regulated power is provided to the RF transmitter load. The first stage of the power conversion, which is AC-to-DC power conversion, is well known and straightforward, so it is not elaborated further. The second stage of the PPU is the energy storage and supply circuit, which has the ability to store electrical energy from the harvester before supplying it to the load. Energy harvested from the piezoelectric wind energy harvester is first stored in a capacitor, and when sufficient energy to power the wireless RF transmitter is accumulated, a triggering signal would trigger the storage circuit to release the stored energy to the RF load. The last stage of the PPU is the voltage regulation stage; the output voltage supplied to the RF transmitter load is regulated by a voltage regulator to be 3.3 V.

The vibration-based piezoelectric wind energy harvester and its related PPU were implemented into a hardware prototype to power the wireless RF transmitter load as shown in Figure 2.42. The physical dimensions of a piezoelectric wind energy harvester (half grey and half white) are 76.7 × 12.7 × 2.2 mm, and it is held tightly at one end by the square clamp (cantilever mounting), and the other end is left to flap freely. When wind blows at the piezoelectric wind energy harvester, AC voltage is generated and converted into DC voltage by the diode bridge rectifier chip. After this, the DC voltage is stored in

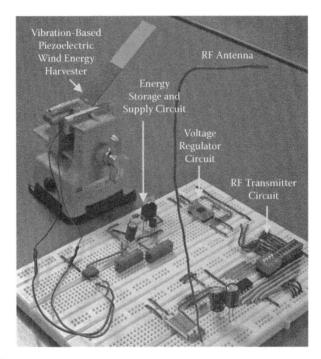

FIGURE 2.42
Photograph of the piezoelectric wind energy harvester system.

a 33-μF capacitor until the preset voltage of around 8.8 V and is subsequently discharged to the RF transmitter load. A detailed schematic diagram of the piezoelectric wind energy harvester system is shown in Figure 2.43. It can be seen from the schematic diagram that the Q1 and Q2 power semiconductor devices residing in the energy storage and supply system act like a control switch that would initiate an on or off signal to the storage capacitor to release the stored energy. The design of the energy storage circuit is adapted from a similar circuit designed by the Massachusetts Institute of Technology (MIT) for digital RFID (RF identification) of piezoelectric [81]. The significant improvement in the revised circuit discussed in this research work is that fewer components have been used in the revised circuit while it can still operate to deliver similar performance. One distinct difference between the proposed design and the MIT design is the turn off process of Q1. In the proposed design, turning Q1 off is determined by the voltage drop across R3 instead of the MIT method of transmitting a negative pulse from MAX666 through C3 to turn off Q1.

Initially, both Q1 and Q2 are off, so the ground lines of the voltage regulator (MAX666) and the RF AM transmitter (AM-RT4-433FR) are disconnected from C1. As C1 charges beyond the preset on voltage threshold of around 8.8 V (the preset voltage level is determined by the zener diode Z1 as 8.2 V and the base emitter junction of Q1 as 0.6 V), the control switch Q1 turns

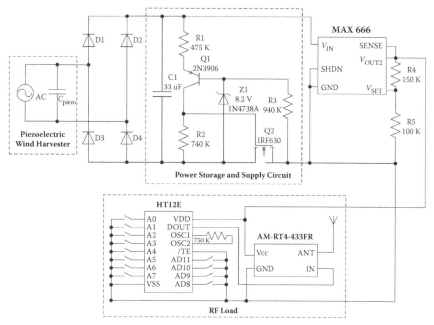

FIGURE 2.43
A schematic diagram of the piezoelectric wind energy harvester system.

on. The moment when Q1 is on, there is a voltage drop across R2 that is higher than the threshold gate-source voltage $V_{gs}(th)$ of Q2 in order to activate the control switch Q2. Once Q2 is activated, Q1 is latched. This connects the ground lines of MAX666 and AM-RT4-433FR with C1, allowing C1 to discharge through the circuitry. MAX666 is a low-power series voltage regulator, which produces a stable +3.3 V for the serial ID encoder (HT12E) and the RF AM transmitter (AM-RT4-433FR) throughout the discharge of C1. When the voltage across C1 drops below the off voltage threshold of around 4.58 V, the voltage drop across R3 causes Q1 to turn off and hence in turn deactivates Q2 from the latched stage. When this happens, the ground lines of MAX666 and AM-RT4-433FR are disconnected from C1, and the discharge of C1 is stopped. Subsequent wind flow through the piezoelectric wind energy harvester increases the voltage on C1, allowing the cycle to start afresh.

2.2.4 Experimental Results

In the experiments, the performances of the piezoelectric wind energy harvester system are explored and evaluated based on a wind speed of 6.7 m/s, which is the preset threshold wind speed in Figure 2.40 to trigger the RF wireless transmitter. Whenever the wind speed reaches the preset wind speed of 6.7 m/s, the electrical energy stored in the capacitor is supplied to the RF transmitter. The encoded digital information is then transmitted to the base

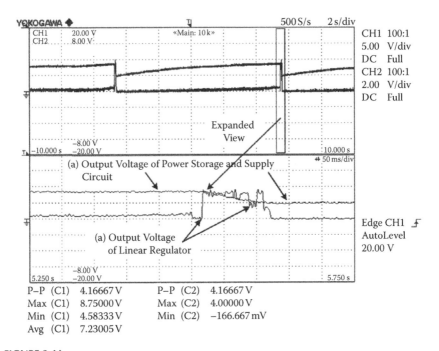

FIGURE 2.44
Waveforms of (a) a charging and discharging of the output voltage of energy storage and supply circuit and (b) the output voltage of voltage regulator.

station in a wireless manner. Referring to Figure 2.44, it can be seen that the output voltage of the harvester takes about 10 s to charge the storage capacitor to maximum voltage V_{max} of around 8.8 V. As soon as the capacitor reaches V_{max}, the electrical energy stored in the capacitor is discharged to the RF load, and the voltage across the capacitor starts to decrease to the minimum voltage V_{min} of 4.58 V in 100 ms. After this, the cycle starts again.

To find the amount of energy that has been stored in the storage capacitor during the time when the capacitor is charged from V_{min} to V_{max}, Equation 2.33 is applied. The energy stored in the capacitor is calculated to be 917 μJ.

$$E_{cap} = \frac{1}{2}C\left(V_{max}^2 - V_{min}^2\right) \tag{2.33}$$

The electrical power consumed by the RF transmitter load is dependent on the number of digital-encoded data word being transmitted. For each data word, the time taken for one transmission is 20 ms, that is, 10 ms of active time and 10 ms of idle time. During the active transmission time, the RF load requires a supply voltage and current of 3.3 V and 4 mA, respectively, to operate. As for the remaining time of 10 ms, the RF load is operating in idle mode,

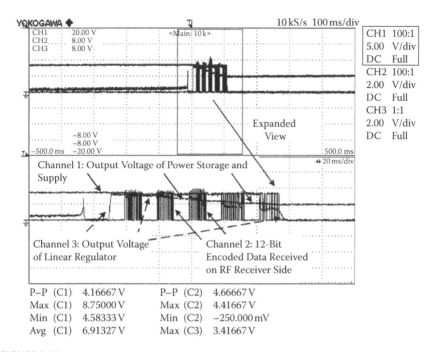

FIGURE 2.45
Waveforms collected at the RF receiver side to the display number of encoded data words received using the harvested energy.

which means that a very minimal amount of energy would be consumed, so it is reasonable to exclude the power being consumed by the RF load during the idle time. By calculation, the average power and hence the energy consumed by the RF transmitter are 13.2 mW and 132 μJ, respectively. Taking into consideration the power loss in the voltage regulator, the total power required for one digital 12-bit data word is 167 μJ. Based on the harvested electrical energy of 917 μJ stored in the capacitor, five packets of 12-bit digital-encoded data words (each packet consumes 167 μJ) can be transmitted. This is verified by the five packets of digital-encoded data words received at the base station shown in Figure 2.45. Referring to Figure 2.45, channels 1 and 3 exhibit the output voltages of the storage capacitor and the voltage regulator; channel 2 shows the encoded data words received at the RF receiver end.

For the transmission of five packets of 12-bit encoded digital data words, 835 μJ of electrical energy is drawn from the bank (917 μJ) of electrical energy stored in the capacitor. Of the 835 μJ of electrical energy drawn, the actual useful energy being consumed by the RF transmitter is 660 μJ. Therefore, the efficiency of the system is around 72%. Power loss is incurred in the power management circuit. In addition, it can be observed in Figure 2.45 that during the discharge period of 100 ms, as the capacitor voltage is decreasing from

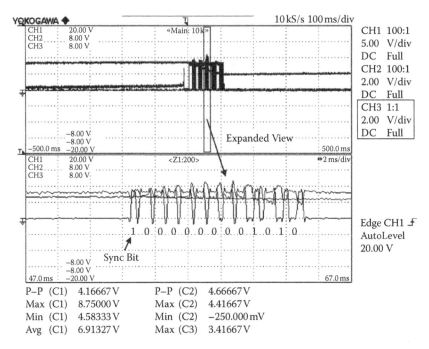

FIGURE 2.46
Waveforms collected at the RF receiver side to verify the 12-bit encoded data word.

8.75 to 4.58 V, the voltage regulator is able to maintain its output voltage at 3.3 V for around 90 ms. After this, the output voltage of the voltage regulator starts to decay with the voltage across the storage capacitor. Figure 2.46 shows the waveform of a 12-bit decoded data word that is generated by the HT12D decoder at the receiver end. The decoded data sequence of the waveform is read as [0000 0000 1010]. Comparing the decoded data sequence with the encoder data sequence at the transmitter side (starting with a synchronization bit following an 8-bit address [0000 0000] and 4-bit data [1010]), the two data sequences match each other quite accurately. The positive outcome concluded from the experimental results verify that the RF transmitter is able to successfully transmit five encoded digital data words to the RF receiver at the base station in a wireless manner using the wind energy harvested by the proposed piezoelectric wind energy harvester system.

2.2.5 Summary

A novel way of harvesting wind energy using piezoelectric material for a low-power autonomous wind speed sensor has been proposed and demonstrated with an experimental setup. Unlike the conventional wind turbine, the piezoelectric wind harvester together with its related electronic circuits are relatively smaller in size and lighter in weight; hence, it is more portable and suitable for low-power autonomous sensors. Experimental results showed that the designed piezoelectric wind harvester system with an efficiency of 72% is able to power a wireless RF transmitter load to transmit five digital 12-bit encoded data words to the receiver end during one wireless transmission. The proposed piezoelectric-based wind harvester can be used in applications that need to detect wind speed beyond a certain threshold level, for instance for early warning of a storm. When the ambient wind speed is beyond the preset value, a 12-bit digital signal will be transmitted in a wireless manner to the base station to set off the warning siren.

3

Thermal Energy Harvesting System

Thermal energy harvesting (TEH) is the process of converting thermal energy to electrical energy by using a thermoelectric generator (TEG) made of thermocouples. Thermoelectric power generators have been successfully developed for decades for kilowatt-scale power generation by using waste heat from industrial processes such as vehicle exhaust, space travel, and so on [84]. Such systems involve heat flows at the kilowatt scale and a temperature of hundreds of degrees Centigrade. However, small-scale TEGs [85–86] for obtaining power on the order of milliwatts or lower from ambient thermal energy sources with small temperature differences have only recently been researched [37]. The challenges with utilizing TEG technology in small-scale TEH are low energy conversion efficiency, inconsistency, and low output power due to temperature fluctuation and high costs [85]. As such, there is a significant need for an efficient power management circuit to maximize the power transfer from the TEG source to its connected electronic load over a wide range of operating conditions.

For decades, maximum power point tracking (MPPT) schemes [87–90] have been proposed for large-scale power generation systems. However, these MPPT schemes are not suitable for small-scale energy harvesting systems as they consume a significant amount of power for their continuous operation. At lower power levels of milliwatts of interest in this chapter, implementation of such accurate MPPT schemes for small-scale TEH, whereby the power consumed by the complex MPPT circuitry could be higher than the harvested power itself, is not desirable. It is thus important to ensure that the gain in input energy is always higher than the additional losses that are caused by the MPPT operation. Thus far, limited research can be found in the literature that discusses a simple but yet compatible MPPT algorithm addressing the issue of a small-scale TEH system. This chapter presents a resistor emulation approach and its associated circuitry for harvesting near maximum energy from the thermal energy source. The rationale behind the resistor emulation approach [69–71] is that the effective load resistance is controlled to emulate the source resistance of the thermoelectric generator to achieve impedance matching between the source and load; hence, the harvested power is always at its maximum at any operating temperature difference. A power-electronic-based converter with minimal open-loop control overhead is realized to act as a near-constant resistance at its input port to emulate the TEH source while transferring energy to the time-varying load resistance, which consists of an energy storage element and a wireless sensor node.

The rest of the chapter is organized as follows: Section 3.1 describes the TEH system and energy conversion effect. Section 3.2 discusses the resistor emulation-based maximum power point (MPP) tracker for TEH; Section 3.3 describes its implementation using a DC-DC (direct-current-to-direct-current) buck converter. The experimental results of the optimized TEH wireless sensor node prototype are discussed in Section 3.4, followed by a summary of the TEH research work in Section 3.5.

3.1 Thermal Energy Harvester

The thermal energy harvester employed in this chapter for converting thermal energy into electrical energy is shown in Figure 3.1. The thermal energy harvester is designed for two main purposes: (1) to house the miniaturized thermoelectric generator for ease of deployment and (2) to channel the thermal energy, generated from the heat source at a certain high temperature of T_H, to the enclosed TEG via a thin film of thermally and electrically conductive silver grease between them and then release the residual heat accumulated in the heat sink to the surrounded ambient air at a lower temperature T_C.

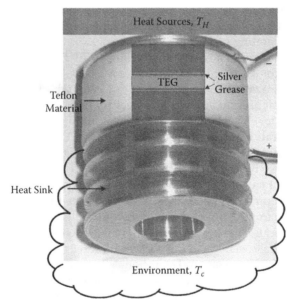

FIGURE 3.1
A thermal energy harvester consisting of a housing structure and a TEG.

3.1.1 Description of a Thermoelectric Generator

Thermoelectricity describes the relationship between heat flow and electrical potential in conducting materials. The ability to generate electrical power from a temperature gradient in materials is due to Seebeck's effect [91]. This Seebeck's effect can be observed in a thermocouple made of two dissimilar conductors. If the two junctions are maintained at different temperatures, that is, T_{CJ} and T_{HJ}, an open-circuit voltage proportional to the temperature difference ΔT_{TEG} would be developed. For a TEG, which is composed of n thermocouples connected electrically in series and thermally in parallel, the open-circuit voltage V_{oc} of the TEG is given as

$$V_{oc} = S * \Delta T_{TEG} = n * \alpha(T_{HJ} - T_{CJ}) \tag{3.1}$$

where α and S represent the Seebeck's coefficient of a thermocouple and a TEG, respectively. In this TEH research, a Thermo Life TEG is used; it is based on the development of a unique thin-film technology for the deposition of highly efficient thermoelectric materials of the Bi_2Te_3 type on thin Kapton foils [91]. Based on the technical datasheet provided by Thermo Life, the essential geometrical and thermal parameters of the TEG prototype are presented in Table 3.1. The high output voltage of the Thermo Life TEG, due to over 5200 thermocouples in series as well as the relatively high thermal resistance, makes this device ideal for energy scavenging, where only a small temperature gradient is available, such as in body and waste heat applications [91].

3.1.2 Analysis of the Thermal Energy Harvester

To illustrate the thermal and electrical characteristics of the thermal energy harvester, an equivalent electrical circuit model is provided in Figure 3.2.

TABLE 3.1

Main Parameters of Thermo Life Prototype

Geometrical Parameter	Unit	Value
Total height of device	mm	0.85 ± 0.05
Outer diameter of device	mm	9.60 ± 0.05
Volume of device	mm^3	62 ± 5
Mass of device	mg	185 ± 2
Number of thermocouples (n)		5200
Thermal Parameter		
Total thermal resistance (R_{TEG})	K/W	14.1 ± 1.0
Seebeck's coefficient of TEG (S)	V/K	1.1
Seebeck's coefficient of a thermocouple (α)	mV/K	0.21

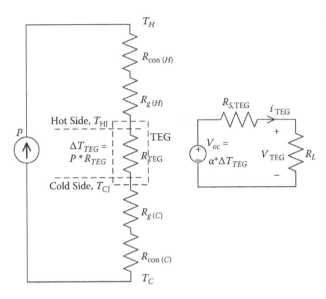

FIGURE 3.2
An equivalent electrical circuit of the thermal energy harvester.

3.1.2.1 Thermal Analysis

Referring to Figure 3.2, it can be observed that the TEG is connected to the hot and cold reservoirs via the thermal contact and thermal grease resistances, which are given by $R_{con(H)}$, $R_{g(H)}$ and $R_{g(C)}$, and $R_{con(C)}$, respectively. Considering all these thermal resistances R_{Total} (see Figure 3.2, left circuit) residing in the housing structure of the thermal energy harvester and comparing them with the TEG's internal thermal resistance R_{TEG}, the actual temperature drop across the thermoelectric generator ΔT_{TEG} may then be expressed as

$$\Delta T_{TEG} = \Delta T * \frac{R_{TEG}}{R_{Total}}$$

$$= (T_H - T_C) * \left[\frac{R_{TEG}}{R_{con(H)} + R_{g(H)} + R_{TEG} + R_{g(C)} + R_{con(C)}} \right] \quad (3.2)$$

Due to the finite thermal resistances of the thermal energy harvester, the temperature difference ΔT_{TEG} across the junctions of the TEG is lower than the temperature gradient ΔT that is externally imposed across the thermal energy harvester. To minimize this negative effect, the thermal resistance R_{TEG} of the TEG must be as high as possible, or in other words, the rest of the thermal resistances of the thermal energy harvester must be minimized.

The unwanted thermal resistance of the thermal energy harvester, which is defined as $R_{Thermal} = \Delta x / K A$, can be minimized by carrying out some appropriate hardware design on the thermal energy harvester, such as: (1) increasing the contact surface A of the heat transfer area, (2) reducing the thickness of the material Δx used like the fins of the heat sink, and (3) selecting aluminium material, which has very high thermal conductivity K for good thermal conduction. To maintain the temperature difference across the thermal energy harvester, the Teflon material is inserted between the hot and cold sides of the housing structure shown in Figure 3.1 to prevent the adverse effect of parallel parasitic thermal resistance [92].

3.1.2.2 Electrical Analysis

It should be noted that the effective temperature gradient ΔT_{TEG} depends not only on the thermal and electrical properties of the TEG but also on the resistive load that is connected. When connecting a load resistance R_L to the TEG as shown in Figure 3.2, a current I_{TEG} flows, which is given by Dalola et al. [93].

$$I_{TEG} = \frac{V_{oc}}{R_{s,TEG} + R_L} = \frac{S * \Delta T}{R_{s,TEG} + R_L} \tag{3.3}$$

Depending on the dimension of the TEG, that is, h and A_{leg}, which are the height and the area of a single thermocouple leg, respectively, and the electrical resistivity ρ of the materials used, the TEG internal electrical resistance $R_{s,TEG}$, which is composed of n thermocouples of p-type and n-type semiconductor materials, is given by Dalola et al. [93].

$$R_{s,TEG} = n * 2 * \frac{\rho h}{A_{leg}} \tag{3.4}$$

The output power P_L delivered by the TEG to the load R_L can be expressed as

$$P_L = I_{TEG}^2 R_L$$
$$= S^2 \Delta T_{TEG}^2 \frac{R_L}{(R_{s,TEG} + R_L)^2} \tag{3.5}$$

Referring to Equation 3.5, it can be seen that the output power P_L is dependent on both the TEG electrical resistance $R_{s,TEG}$ and the electrical resistance of the external load R_l. Under an impedance-matching condition where the load resistance R_L is equal to the internal electrical resistance $R_{s,TEG}$, the TEG is generating the maximum output power given by

$$P_{L,MPPT} = \frac{S^2 \Delta T_{TEG}^2}{4 R_{s,TEG}} \tag{3.6}$$

During operation of the TEG, the output voltage is reduced by the ohmic voltage drop across its internal resistance $R_{s,TEG}$. Consequently, the voltage at its maximum power V_{MPPT} is about half that of the open-circuit voltage V_{oc} ($V_{oc} = S\Delta T \approx 2V_{MPPT}$), and the maximum power changes with temperature difference ΔT^2.

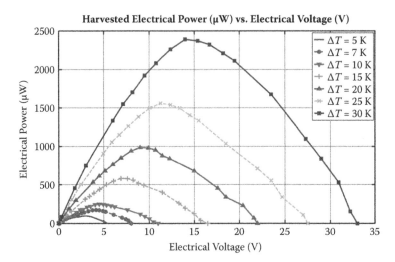

FIGURE 3.3
P-R curves of thermoelectric generator at different thermal gradients.

3.1.3 Characterization of a Thermal Energy Harvester

Based on the thermal analysis, the thermal energy harvester has been designed to maximize the overall output power of the TEG. The geometric design parameters, such as Δx and A, and the thermal interface-related parameters are the main design factors of the thermal energy harvester. The physical size of the optimized thermal energy harvester prototype is 20 x 20 x 20 mm. Some characterization works are performed by applying a temperature difference $(T_H - T_C)$ between the energy harvester faces and measuring both the output voltage and the current with different loads connected. This operation is repeated for temperature differences in the range between 5 and 30 K.

Referring to the power curve (power versus load voltage) shown in Figure 3.4, it can be observed that the maximum obtainable power for each thermal gradient corresponds to an output voltage of the thermal energy harvester. This is unlike the case of the other energy harvesting sources like solar, vibration, and so on, where their power curves peak near a particular output voltage of the energy source. As such, it is not possible to utilize this simple and ultralow-power MPPT approach, that is, fixed reference voltage to the thermal energy harvester. In order to overcome that, Kim et al. [90] propose the adaptive and tracking MPPT approaches that are suitable for TEH; however, these energy-hungry approaches require high computational power and cost with respect to the milliwatt or even lower harvested power levels of interest in this chapter.

Since most of the conventional MPPT approaches are not suitable for TEH, this chapter presents an alternative MPPT technique based on the concept of emulating the load impedance to match the source impedance as illustrated by the power curves in Figure 3.3. This technique is also known as the resistor emulation or impedance matching approach. The power curve plotted in

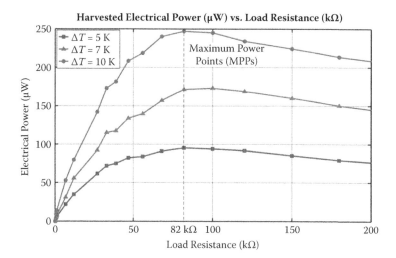

FIGURE 3.4
P-V curves of a thermoelectric generator at different thermal gradients.

Figure 3.4 shows that when the load resistance matches the source resistance of the thermal energy harvester of 82 kΩ, the harvested power is always maximum for different temperature differences. However, for other loading conditions, shifting away from the internal resistance of the thermal energy harvester, either very light or heavy electrical loads, the electrical output power being generated by the generator drops significantly. This exhibits that the MPPT technique based on resistor emulation is a possible option to assist the small-scale TEH system to achieve maximum power harvesting from the thermal energy harvester.

3.2 Resistor Emulation-Based Maximum Power Point Tracker

Resistor emulation techniques have been widely used in the impedance-matching applications [69–71]. For instance, Paing et al. [70] successfully demonstrated the resistor emulation approach in energy harvesting from variable low-power radiative radio-frequency (RF) sources. Khouzam and Khouzam [69] have also discussed the resistor emulation concept used in their direct-coupling approach for optimum impedance matching between the energy harvester and its load by carefully selecting the harvester's rated parameters with respect to the load parameters. However, very limited research works can be found in the literature that discuss the design and implementation of resistance emulation techniques and their approximation for a small-scale TEH system to achieve resistor emulation-based MPPT.

In this chapter, the proposed MPP tracker uses a power converter as an open-loop resistance emulator to naturally track the MPP of the thermal

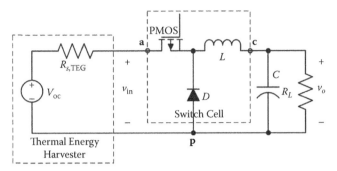

FIGURE 3.5
Buck converter.

energy harvester with very little control circuit overhead. A buck converter topology is selected for the power converter due to the high output voltage of 5 to 35 V (see Figure 3.4) generated by the thermal energy harvester (input voltage V_{in} to the buck converter). The main purpose of the buck converter is to match the optimal resistance of the thermal energy harvester, that is, $R_{s,TEG} = R_{opt} = 82$ kΩ at the converter input port, and efficiently transfer the energy to its output port based on the voltage and charge characteristics of the energy storage element. A previous approach [70] has shown that over a certain range of input power level, operation of a boost (step-up) converter in discontinuous conduction mode (DCM) with a fixed duty cycle results in maximum output power. The results as reported by Paing et al. [70] and Paing and Zane [94] can be related to this research work by showing that the DC-DC converters (e.g., buck, boost, and buck-boost) in DCM acts as a near-constant resistance at its input port for large step-up/step-down conversion ratio.

To fully understand how the buck converter, when operating in DCM, emulates the source resistance of the thermal energy harvester ($R_{s,TEG} = 82$ kΩ) to achieve MPPT, the electrical model of the buck converter shown in Figure 3.5 has been modelled into an averaged equivalent circuit model as shown in Figure 3.6. The modelling process is based on the analysis made by Erickson et al. [71] that shows the average voltage and current of the semiconductor switch are proportional, thus obeying Ohm's law, and the switch can be

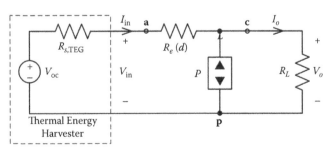

FIGURE 3.6
An averaged equivalent circuit of a buck converter.

simply replaced by an effective resistor, $R_e(d)$. In a similar context, the average diode's voltage and current obey a power source P characteristic, with power equal to the power effectively dissipated by $R_e(d)$. Hence, the buck converter in Figure 3.5 can be represented by the averaged equivalent circuit shown in Figure 3.6, where the switch and diode are replaced with an effective resistor $R_e(d)$ and a dependent power source P, respectively [71].

Based on the DC analysis of the buck converter in DCM [95], the conversion ratio M of the buck converter can be derived [96] as follows:

$$M = \frac{V_o}{V_{in}} = \frac{2}{1 + \sqrt{1 + 4R_e(d)/R_L}} \tag{3.7}$$

In addition, the emulated resistance of the buck converter in DCM $R_e(d)$ is obtained as follows [71]:

$$R_e(d) = \frac{2L}{d^2 T_s} \tag{3.8}$$

where d and $T_s = 1/f_s$ are the duty cycle and the switching period/frequency of the gating signal of the PWM (pulse width modulation) switch, respectively. In practical cases, the step-down conversion ratio M of the buck converter ($V_{in} \gg V_o$) may not be large enough such that the correction factor $1/(1-M)$ is approximately unity. As such, taking into account the effect of the correction factor M on the emulated resistance $R_e(d)$ equation of the buck converter derived in Equation 3.8, the overall emulated resistance R_{em} then becomes

$$R_{em} = R_e(d) \left(\frac{1}{1-M} \right)$$
$$= \frac{2Lf_s}{d^2} \left(\frac{1}{1 - \left(\frac{2}{1+\sqrt{1+8Lf_s/d^2 R_L}} \right)} \right) \tag{3.9}$$

Knowing that the source resistance of the thermal energy harvester $R_{s,TEG}$ as obtained experimentally in Figure 3.3 is fixed at 82 kΩ, the input port of the buck converter has to constantly emulate this resistance value such that MPPT can be achieved. In this DCM operation, the converter parameters such as L and f_s are fixed, whereas the duty cycle d is tuned to the required value to achieve an equivalent emulated resistance. Hence, the equation expressed in Equation 3.9 has been resolved to determine the duty cycle d_{em} that can emulate the source resistance of the thermal energy harvester ($R_{s,TEG} = 82$ kΩ) to achieve MPPT, which is given by Dalola et al.

$$d_{em} = \sqrt{\frac{2Lf_s}{R_{em} \pm \sqrt{R_{em} * R_L}}} \tag{3.10}$$

To demonstrate the competency of the buck converter as a resistor emulation-based MPP tracker and to verify its governing equation expressed

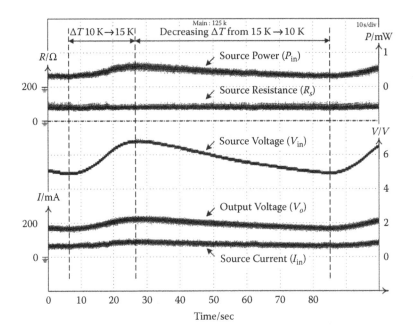

FIGURE 3.7
Operation of a resistor emulation-based MPP tracker under varying temperature differences.

in Equation 3.10, the buck converter has been experimentally tested under varying temperature differences and different loading conditions. First, a fixed resistance of 10 kΩ was used as the test load for the operation of a resistor emulation-based MPP tracker under varying temperature differences; the experimental results are shown in Figure 3.7.

Referring to Figure 3.7, as the temperature difference ΔT across the thermal energy harvester increases from 10 to 15 K, the harvested power P_{in}, which is the result of the product of the source voltage V_{in} and source current I_{in}, increases from around 300 to 600 μW. During this time when ΔT is increasing, it can be seen from Figure 3.7 that the source resistance R_s is emulated by the resistor emulation MPP tracker to remain steadily at the optimal resistance of 82 kΩ.

Based on Seebeck's effect, as the temperature difference across the thermal energy harvester rises from 10 to 15 K, the source voltage increases from 5 to 7 V, and the step-down voltage V_o output from the buck converter with resistor emulation-based MPP tracker also increases from 1.8 to 2.2 V. Given that the converter parameters are chosen as $L = 30$ mH, $f_s = 18$ kHz, and $d_{em} = 0.14$, using Equations 3.7, 3.8, and 3.9, the source resistance obtained experimentally, as shown in Figure 3.7, is verified to be around 82 kΩ. The buck converter is able to maintain the thermal energy harvester near its MPPs for various input operating conditions. Hence, the buck converter yields good performance as a simple resistor emulation-based MPPT under

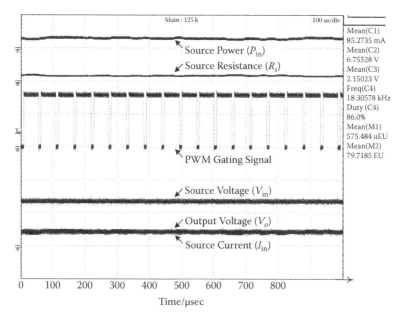

FIGURE 3.8
Operation of a resistor emulation-based MPP tracker at 10-kΩ loading.

varying temperature differences. These observations and analyzes apply to the decreasing temperature difference condition as well.

Second, the operation of a resistor emulation-based MPP tracker for different loading conditions were experimentally tested, and the experimental results are shown in Figures 3.8 and 3.9. Equation 3.10 is used to determine the duty cycle that emulates the source resistance of the thermal energy harvester ($R_{s,TEG} = 82$ kΩ) for maximum power transfer between the source and the load. The operating range of the loading conditions is determined based on the two extreme ends of the operational bandwidth of the buck converter when connected to a 0.1-F supercapacitor, the initial charging and final charged states of the supercapacitor can be emulated by two load resistances R_L of 10 and 56 kΩ, respectively. Based on Equation 3.10, the emulated duty cycles of the two resistance values (i.e., 10 and 56 kΩ) are calculated to be 0.14 and 0.27, respectively.

Referring to Figures 3.8 and 3.9, the complementary emulated duty cycles (a P-MOSFET [metal-oxide-semiconductor-field-effect-transistor] [PMOS] is used instead of an N-MOSFET [NMOS] as the high-side switch of the buck converter) obtained experimentally are 0.14 and 0.251 for the load resistances R_L of 10 and 56 kΩ, respectively, and the experimental results tally with the calculated duty cycles of 0.14 and 0.27. At the emulated duty cycles, it can be observed in both Figures 3.8 and 3.9 that the approximate desired $R_{em} = 82$ kΩ is achieved, and the maximum electrical power $P_{MPPT} = 580$ μW (see Figure 3.3) is harvested.

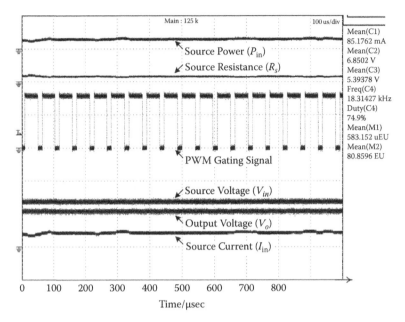

FIGURE 3.9
Operation of a resistor emulation-based MPP tracker at 56-kΩ loading.

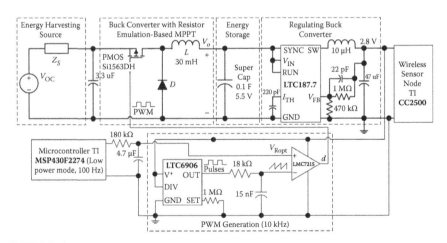

FIGURE 3.10
A schematic diagram of the TEH system.

3.3 Implementation of an Optimal Thermal Energy Harvesting Wireless Sensor Node

The schematic diagram of a self-autonomous wireless sensor node powered by the designed TEH system and its ultralow-power and efficient power management circuit is illustrated in Figure 3.10. Referring to Figure 3.10, the designed power management circuitry with resistor emulation MPPT approach can essentially consist of three main building blocks: (1) a buck converter with MPP tracker and its control and PWM generation circuit that manipulates the operating point of the TEH to keep harvesting power at the MPP, (2) an energy storage element (i.e., supercapacitor to buffer the energy transfer between the source and the load), and (3) a regulating buck converter to provide constant voltage to the wireless sensor node and other electronic circuitries.

3.3.1 Buck Converter with Resistor Emulation-Based Maximum Power Point Tracking

The buck converter in the DCM has been illustrated to perform well as a near-constant resistance at its input port to manipulate the thermal energy harvester to transfer maximum power to its output port, which is connected to a 0.1-F supercapacitor, based on the duty cycle d of the PWM gate signal applied to the buck converter [70]. Within the operational bandwidth of the buck converter under different loading conditions, in the range of 10 to 56 kΩ, a duty cycle value of 0.16 has been selected to operate the buck converter to its near MPPs. Substituting the duty cycle value, $d = 0.16$, into Equation 3.9, the emulated resistances, with the connected load of 10 and 56 kΩ, are calculated to be 68 and 126 kΩ, respectively. Referring to Figure 3.3, it can be observed that the power losses at these resistance values, which are near to the MPPs, are less than 5% of the maximum obtainable power at $R_{em} = 82$ kΩ. This power waste (few tens of microwatts) is much less than the power consumed by the high-power overhead of complex control circuitry required to perform the accurate and precise closed-loop MPPT techniques.

The operation of the buck converter as an open-loop resistance emulator is as follows: A low-frequency PWM control signal, about 100 Hz, of the desired duty cycle of 0.16 is generated by a Texas Instruments microcontroller (TI MSP430F2274). An ultralow-power PWM generation circuit is designed to transform the low-frequency PWM control signal generated from the reduced clock speed microcontroller to a higher switching frequency of 10 kHz, so smaller filter components are used in the buck converter to miniaturize the overall TEH system. The PWM generation circuit is made up of a micro-power resistor set oscillator (LTC6906) used for sawtooth generation and a micropower, rail-to-rail complementary metal-oxide semiconductor (CMOS) comparator (LMC7215), which compares its reference signal ($d_{em} * V_{ref} = 0.16 * 2.5V = 400$ mV) with the high-frequency sawtooth signal to generate the PWM gate signal to control the buck converter.

3.3.2 Energy Storage

For long-term deployment of the TEH wireless sensor node, it is required to have an energy storage device such as supercapacitor and batteries on-board the sensor node to accumulate the input energy generated from the heat source and use it to sustain the node's operation throughout its lifetime. The supercapacitor has superior characteristics over batteries, which include numerous full-charge cycles (more than half-a-million charge cycles), long lifetime (10 to 20 years operational lifetime), and high power density (an order of magnitude higher continuous current than a battery) [34]. Unlike the discrete capacitors, which have very small capacitance values in the picofarad to microfarad range, the supercapacitor has a very large capacitance value in the farad range suitable for energy storage purposes. For the design of the buck converter as a resistor emulation-based MPP tracker, it is important to consider the dynamic response of the supercapacitor to ensure constant MPPT operation is achieved.

For a time period of 500 s as shown in Figure 3.11, the supercapacitor is charged by the TEH system from its initial state of 1 V. As the supercapacitor is charging, the dynamic response of the supercapacitor changes according to the operating conditions of the TEH system; the impedance of the

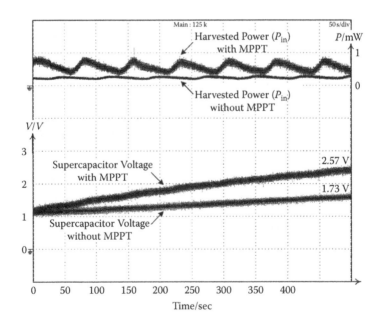

FIGURE 3.11
Performance of a TEH system with MPPT and without MPPT for charging a supercapacitor.

supercapacitor is equivalent to the range of load resistances R_L between 10 to 56 kΩ. At V_{cap} (500 s), the 0.1-F supercapacitor is being charged to voltage levels of 2.57 V with the MPPT scheme and 1.73 V without the MPPT scheme. Comparing the two schemes, it is obvious that the charging performed by the TEH system with MPPT is much higher than its counterpart without the MPPT scheme. This is because the thermal energy harvester is maintained at its near MPPs; hence, more electrical power is transferred to the supercapacitor. When the TEH system is operating with the MPPT scheme, the amount of energy accumulated in the supercapacitor after 500 s is 0.28 J, which is two times its counterpart of 0.15 J; hence, this exhibits the superior performance of the TEH system with the MPPT scheme over its counterpart under a dynamic loading condition.

3.3.3 Regulating a Buck Converter and Wireless Sensor Node

The TEH system is designed to power a commercially available wireless sensor node supplied by Texas Instrument known as the wireless target board, eZ430-RF2500T. A buck converter (LTC1877) obtained from Linear Technology is inserted between the supercapacitor and the wireless sensor node to provide a constant operating voltage of 2.8 V_{DC}. The efficiency of the buck converter is around 80% to 90%, consuming an operating current of 12 µA. In this chapter, the operation of the wireless sensor node deployed in an application field is illustrated in Figure 3.12.

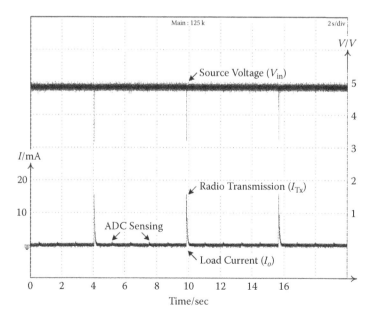

FIGURE 3.12
Operation of the wireless sensor node.

As seen in Figure 3.12, the sensor node's operation is comprised of (1) sensing some external analog signals of sensors such as temperature and (2) communicating and relaying the sensed information to the gateway node every 5 s. On receiving the data at the base station, the collected data is then postprocessed into usable information for any follow-up action. This duty-cycling approach could significantly reduce the power consumption of the energy-hungry radio module of the sensor node with a slower transmission rate of every few seconds.

3.4 Experimental Results

The optimal TEH wireless sensor node has been successfully implemented in a hardware prototype for laboratory testing. Several experimental tests have been conducted to differentiate the performance of the TEH system and its resistor emulation MPPT scheme in powering the connected load consisting of a supercapacitor, control and PWM generation circuitries, and wireless sensor node. The operation of the electrical load, as shown in Figure 3.13, is first powered solely by its onboard supercapacitor and then the TEH system with its integrated MPPT harvesting at a temperature difference of 20 K.

Referring to Figure 3.13, it can be observed that the supercapacitor voltage V_o keeps decreasing during the period of time when neither the TEH system

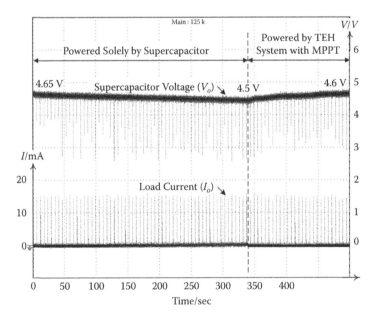

FIGURE 3.13
Operation of a wireless sensor node.

nor its integrated MPPT is connected to the sensor node. The conventional sensor node, which operates solely on the supercapacitor, consumes 200 μW of average power from the 0.1-F supercapacitor; hence, the voltage across the supercapacitor drops from 4.65 to 4.5 V in around 330 s, which is calculated to be 69 mJ of energy transferred to the load. Once the TEH system with MPPT is activated, the harvested power of around 450 μW is used to power the sensor node as well as to charge the supercapacitor back to 4.6 V in 170 s. This indicates that solely depending on the energy storage to sustain the operation of the sensor node is not sufficient; it is only when the TEH system with its MPP tracker is employed that sufficient power is provided for both the operation of the wireless sensor node and charging the supercapacitor.

The designed buck converter with the resistor emulation MPPT approach has already been demonstrated to yield good performance in extracting maximum power from the thermal energy harvester, but this comes at the expense of additional power losses in the converter and its associated control and PWM generation circuits. It is thus necessary to investigate the significance of these power losses as compared to the total harvested power. The first investigation is to determine the efficiency of the buck converter η_{conv} as a function of its output load power P_{load} over its input DC power P_{dc} under different temperature differences and loading conditions. Take, for example, at a temperature difference and output load resistance of 20°C and 10 kΩ, respectively, the efficiency of the buck converter is given by

$$\eta_{conv} = \frac{P_{out}}{P_{in}} * 100\% \tag{3.11}$$
$$= \frac{2.7V^2/10k\Omega}{9V * 88 \ \mu A} * 100\% = 92\%$$

For all other temperature differences and loading conditions, the efficiencies of the buck converter are calculated using Equation 3.11 to be on an average of 90%, and the computed results are shown in Figure 3.14. This high-efficiency buck converter is very favourable and desirable in a very low power rating condition of milliwatts or even lower. Another investigation being carried out is to determine the power consumption of the associated control and PWM generation electronic circuits and its significance as compared to the harvested power. Based on the voltage and current requirements of each individual component in the TEH system shown in Figure 3.10, the total power consumption of the electronic circuits can be calculated as follows:

$$P_{consumed@20K,10k\Omega} = P_{PMOS@V_{MPPT}} + P_{comparator@V_{MPPT}} + P_{filter@2.8V}$$
$$+ P_{oscillator@2.8V} \tag{3.12}$$
$$= 9V * (4 \ \mu A + 21 \ \mu A) + 2.8V * (3 \ \mu A + 20 \ \mu A)$$
$$= 289 \ \mu W$$

These power losses associated with the resistor emulation-based MPP tracker at different temperature differences are illustrated in Figure 3.15. Taking

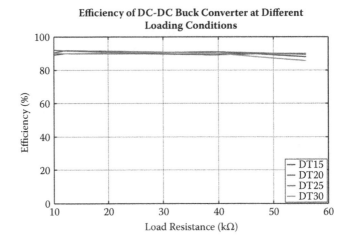

FIGURE 3.14
Efficiency of a buck converter with resistor emulation-based MPPT for various temperature differences.

into account both the power loss across the buck converter and the power loss in the associated control and PWM generation circuits mentioned by Equations 3.11 and 3.12, respectively, the performance comparison between the TEH system with MPPT and without MPPT are tabulated in the bar chart in Figure 3.16. For all the temperature differences shown in Figure 3.16, it can be observed that the electrical power harvested by the TEH system with MPPT is at least two to three times more than the TEH system without MPPT. Taking into account the power losses of the buck converter and its associated circuitry,

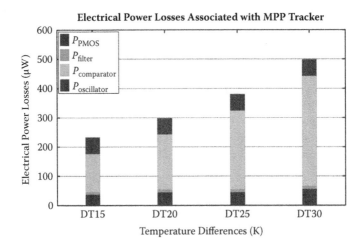

FIGURE 3.15
Power losses associated with an MPP tracker.

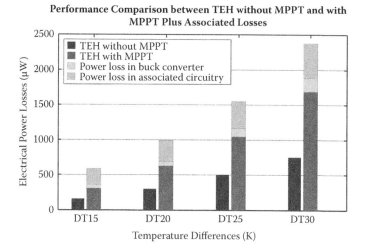

FIGURE 3.16
Performance comparison between the TEH system without MPPT and with MPPT.

the net harvested power from the TEH system with the MPPT scheme is still more than the case without the MPPT scheme. Hence, this exhibits the importance as well as the contribution of implementing MPPT in the TEH system.

3.5 Summary

In this chapter, an efficient TEH system and its power management circuit have been proposed to maximize power transfer from the heat source to its connected wireless sensor node. The electrical characteristic of the TEH system is unlike the case of the other common energy harvesting sources like solar and vibration, where most of the conventional MPPT approaches with very little control overhead are not suitable. This chapter presented a resistor emulation-based MPPT technique to emulate the load impedance to match the impedance of the TEH system. Experimental results showed that the self-autonomous wireless sensor node powered by the designed TEH system and its ultralow-power and efficient power management circuit yielded better performance than the conventional battery-operated wireless sensor node. At a temperature difference of 20 K, the efficiency of the buck converter was 92%, and its associated power management circuit consumed 289 μW of power to operate. From the experimental test results obtained, an average electrical power of 629 μW was harvested by the optimized TEH system at an average temperature difference of 20 K, which is almost two times higher than the conventional energy harvesting method without using MPPT.

4

Vibration Energy Harvesting System

Many environments, such as highways, railways, and so on, are subjected to ambient vibration energy that is not commonly used. To use these ambient vibrations as a power source, many researchers have successfully built and tested three basic methods for generating electrical energy from this vibration energy source: electromagnetic induction [48], electrostatic generation [97], and piezoelectric materials [98]. While each of these techniques can provide a useful amount of energy, piezoelectric materials have received the most attention due to their ability to directly convert applied strain energy into usable electric energy and the ease at which they can be integrated into a system [98]. Unlike the electrostatic and electromagnetic approaches, which require a complex "two-part" design (the two plates of the variable capacitor in the electrostatic configuration, the coil and the magnet in the electromagnetic one), the piezoelectric approach is relatively simpler in design and implementation. Plus, Roundy et al. [99] demonstrated that the piezoelectric type has the highest energy density. Based on these positive findings, the piezoelectric approach has been employed in this vibration energy harvesting (VEH) research for powering the electrical load.

Piezoelectricity is the ability of some materials (i.e., crystals) to convert mechanical energy into electrical energy and the inverse [100]. When an external force mechanically strains the piezoelectric material, the material becomes electrically polarized, and the degree of polarization is proportional to the applied strain. The opposite effect is also possible: When the piezoelectric material is subjected to an external electrical field, it is deformed. The relationships between the applied force and the subsequent response of a piezoelectric material depend on three factors [101]: (1) the dimensions and geometry of the material, (2) the piezoelectric properties of the material, and (3) the directions of the mechanical or electrical excitation. The first relationship is straightforward. As for the second relationship, the behaviour of piezoelectricity can be modelled with the following constituent equations:

$$D = d X + \epsilon^X E \tag{4.1}$$

$$x = s^E X + d E \tag{4.2}$$

Based on the describing electromechanical expression of a piezoelectric material expressed in Equation 4.1, the electrical displacement D relates with

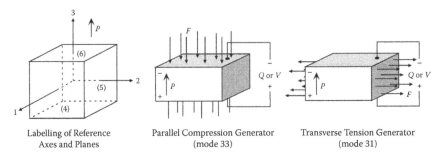

| Labelling of Reference Axes and Planes | Parallel Compression Generator (mode 33) | Transverse Tension Generator (mode 31) |

FIGURE 4.1
Parallel and transverse tension modes of operation for a piezoelectric generator. (From J.L. Gonzalez, A. Rubio, and F. Moll, "Human-powered piezoelectric batteries to supply power of wearables electronic devices," *International Journal of the Society of Materials Engineering for Resources*, vol. 10, no. 1, pp. 34–40, 2002 [79].)

the mechanical stress X applied, and the electrical field E is generated. The proportionality constants are the d coefficient and the dielectric constant ϵ^X measured at constant stress. As for the expression in Equation 4.2, it relates the mechanical strain developed when an electrical field E is applied with the field, and the mechanical stress X is developed. The proportionality constants are the same d coefficient and the elastic compliance s^E measured at a constant electrical field. For the third relationship, according to Gonzalez et al. [79], the piezoelectric materials are visualized as 3D (three-dimensinal) structures where the mechanical and electrical magnitudes can be applied or measured in any of the three axes. The axes are defined in Figure 4.1 where the impact-based piezoelectric generators in parallel and transverse modes of operation are illustrated.

In this VEH research, two types of piezoelectric generators that are designed to harvest impact or impulse forces are explored: (1) the piezoelectric push-button igniter described in Section 4.1 and (2) the prestressed piezoelectric diaphragm material described in Section 4.2. With reference to Figure 4.1, the piezoelectric igniter operates in the parallel compression mode or 33-mode of operation where the electrical field is generated (i.e., axis- 3) across the same axis where an external mechanical force F is applied to create a mechanical resonance in the piezoelectric element [79]. As for the other type of piezoelectric generator, which uses a prestressed piezoelectric diaphragm material, a transverse mode or 31-mode of operation as shown in Figure 4.1 is considered [102]. The mechanical strain applied on the piezoelectric material is perpendicular to its output electrodes, so the surface where the charge is collected and the surface where the force is applied are independent. When the piezoelectric material is excited in 31-mode operation, elongation or compression normal to the 1 axis is induced; hence, an electric field normal to the 3 axis is generated.

Impact-based piezoelectric energy harvesting has been widely discussed in the literature to harvest waste kinetic energy from human motions, ammunitions, and so on for powering low-power electronic devices. According to

Beeby et al. [103], the earliest example of a piezoelectric kinetic energy harvesting system extracted energy from impacts. Umeda et al. [104] pioneered the analysis of the energy generated by the impact of a steel ball on a piezoelectric membrane. Initial work explored the feasibility of this approach by dropping a 5.5-g steel ball bearing from 20 mm onto a piezoelectric transducer. Keawboonchuay et al. [105] studied a high-power impact piezoelectric generator that can be incorporated into ammunitions. Subsequently, impact coupling of a piezoelectric transducer designed for use in human applications was described by Renaud et al. [106]. The authors presented some analysis related to the impact harvester, which was comprised of an inertial mass confined within a frame but free to slide along one axis, for human applications and assessed the relevance of such a system for harvesting energy from large-amplitude and low-frequency excitations. Energy is generated when the sliding mass strikes steel/PZT (lead zirconate titanate) cantilevers located at each end of the frame. Several studies on direct straining of, or impacting on, a piezoelectric element for human applications have also been reported. One of the earliest examples of a shoe-mounted generator incorporated a hydraulic system mounted in the heel and sole of a shoe coupled to cylindrical PZT stacks [107]. The hydraulic system amplifies the force on the piezoelectric stack whilst reducing the stroke. A subsequent device was developed at the Massachusetts Institute of Technology (MIT) in the 1990s [81] as an insole in a sports training shoe; the bending movement of the sole strains both polyvinylidene fluoride (PVDF) stacks to produce electrical power.

4.1 Impact-Based Vibration Energy Harvesting (VEH) Using a Piezoelectric Push-Button Igniter

A self-powered push button is an interesting application that utilizes impact-based energy harvesting [28] for powering a self-powered remote controller. In 1956, Robert Alder designed a self-powered remote control called the Space Commander for Zenith televisions. It featured a set of buttons that struck aluminum rods to produce ultrasonic waves when decoded at the television receiver would change channels appropriately. Joe Paradiso and Mark Feldmeier took this theme further in 2001 by using a piezoelectric element with a resonantly matched transformer and conditioning electronics that, when struck by a button, generate electrical power to run a digital encoder and a radio transmitter [108]. Furthermore, a German company, EnOcean [109], developed some self-powered radio transmitters, energized by a bi-stable piezoelectric cantilever that snaps when pressed, and are conditioned by a switching voltage regulator. In this impact-based VEH research, the proposed piezoelectric push-button igniter system, as depicted in Figure 4.2, is designed to be compact, simple, low cost in terms of power density and energy requirements, and at the same time able to harvest sufficient energy from the impulse force generated by human pressing to power the remote controller.

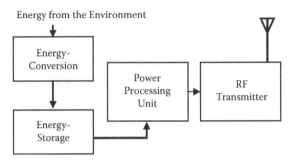

FIGURE 4.2
A block diagram of a self-powered wireless RF transmitter.

Referring to Figure 4.2, the key components of a self-powered wireless transmitter system include a micropower impact-based piezoelectric generator, an energy storage device, a power processing unit (PPU), and a radio transmitter. When a small mechanical force is depressed on the piezoelectric push button, the kinetic energy is harvested by the piezoelectric igniter and converted into electrical energy. The electrical energy is temporarily stored in an energy storage device like a capacitor or battery and then regulated by the PPU into regulated output voltage. After this, the regulated direct current (DC) power provides energy for the radio-frequency (RF) transmitter circuit to perform serial transmission via the radio transmitter to communicate with the environmental sensors, control, and actuating systems deployed in the smart environment.

4.1.1 Piezoelectric Push Button

Piezoelectric push buttons shown in Figure 4.3 have been widely used in industries for several purposes. One example is the integration of a push button in an igniter for gas lighting; it generates very high voltage at very low mechanical impact force, and this high voltage is applied to an air gap to generate an electric arc. Referring to Figure 4.3, the piezoelectric igniter consists of a cylindrical metal base that connects to the negative wire, and the other positive wire is internally connected to the piezoelectric element. A depressible ignition button-like structure is found on the top part of the igniter body. When the piezoelectric push button is depressed, a spring inside will be compressed. When the pressure exceeds a threshold, the spring-loaded hammer will be released, which delivers a dynamic mechanical force to compress the piezoelectric element as illustrated in Figure 4.4. Referring to Figure 4.4, the external force mechanically strains the internal piezoelectric element; these polarized unit cells shift and align in a regular pattern in the crystal lattice. The discrete dipole effects accumulate, developing an electrostatic potential between opposing faces of the element [110].

In the parallel mode of operation [79], the electrical field is generated (i.e., axis-3 in this case as shown in Figure 4.4) across the same axis where the

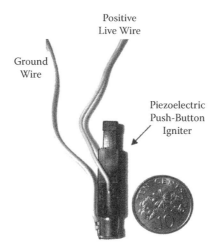

FIGURE 4.3
A piezoelectric push-button igniter and its components.

external mechanical stress, σ, is applied by the spring-loaded hammer. The short circuit charge displacement Q_3 generated on the surface of the piezo-electric generator of area $A = L * W$ along the axis-3 can be expressed as

$$Q_3 = d_{33}F_3 \tag{4.3}$$

where L is the length along the axis-2 and W the width along the axis-1. The applied stress, σ, is a function of the applied mechanical force, F, and the surface area, A. The piezoelectric voltage constant, g, is the quotient of the electric field, generated E, and the stress, T, applied [79]. Additionally, the electric field is a function of the open circuit voltage, V_{oc}, of the piezoelectric generator divided by its thickness, T, hence the open circuit electric field E_{oc} generated [111] can be expressed as

$$E = \frac{V_{oc,3}}{t} = \frac{g_{33}F_3}{A} \tag{4.4}$$

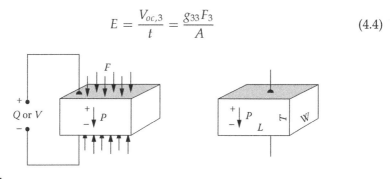

FIGURE 4.4
Parallel compression on a piezoelectric element. (Form Piezo Systems, Inc., "Introduction to piezo transducers," http://www.piezo.com/tech2intropiezotrans.html, accessed on May 17, 2010 [111].)

Knowing that the piezoelectric charge constant, d, [79] is given as

$$d_{33} = \epsilon_{33}^T \epsilon_o g_{33} \tag{4.5}$$

where ϵ_o is the permittivity of free space (8.85 x 10^{-12}) and ϵ_{33}^T is the permittivity of the material at constant stress. It can be demonstrated that the piezoelectric generator can be modelled as a capacitor of value [111],

$$C_3 = \frac{\epsilon_{33}^T \epsilon_o WL}{t} \tag{4.6}$$

Hence, the electrical power P_3 generated by the piezoelectric generator can be calculated as the rate of the energy stored in the capacitor, which is given by

$$P_3 = \frac{1}{2} C_3 V_3^2 f = \frac{1}{2} g_{33} d_{33} F_3^2 \frac{t}{A} f \tag{4.7}$$

where f is the frequency of the vibration. The derived relationship shows that for a given piezoelectric material of fixed area A and thickness T the generated electrical power P under the force F is dependent on the piezoelectric material charge constant d and voltage constant g. In order to obtain a high-performance impact-based piezoelectric generator, the piezoelectric ceramic element with high figures of merit (d and g) has to be selected. In addition, the volume of the piezoelectric element and the amount of stress exerted on the element are also the key factors to be considered in converting mechanical input to electrical energy.

The next step is to determine the electrical characteristic of the piezoelectric push-button igniter to facilitate the design of the PPU for powering the wireless transmitter load. The dimension of the piezoelectric push button used in this research is 35 mm long and 5 mm in diameter and has a deflection of 4.5 mm at a maximum force of 15 N. When an actuation force of 15 N is applied onto the piezoelectric push button [108], the internal hammer is released to strike the piezoelectric element as shown in Figure 4.5. The excitation force of the hammer generates a pressure wave on the piezoelectric element to output a very high voltage of 6 kV. The high output voltage generated follows closely to an alternating current (AC) signal due to dynamic polarization of the piezoelectric element. Each strike of the hammer creates a mechanical resonance in the piezoelectric element as illustrated in the zoomed version of Figure 4.5 whereby 10 small AC voltage pulses are observed. Subsequently, about nine more iterations of the hammer striking the piezoelectric element are observed in Figure 4.5, and the harvested energy pulses from the piezoelectric push-button generator occur for a short time span of around 1.5 ms.

Since each input actuation force provided by the human is not continuous but rather behaves in a pulse-like manner, the output voltage will gradually attenuate following the attenuation of the mechanical vibrations in the

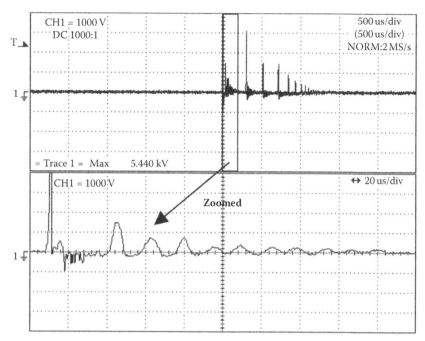

FIGURE 4.5
A piezoelectric push-button igniter.

piezoelectric element. Hence, it is critical to ensure that the harvested energy is sufficient to transmit a few packets of information through the RF transmitter in a wireless manner. Figure 4.6 shows the 12-bit data encoded by a HT-12E encoder to be sent out by the RF wireless transmitter.

The encoder first generates a synchronization starting bit following an 8-bit address [1000 0000] and 4-bit data [0101] sequence serially. The address/data pins of the HT-12E encoder have been prefixed to the [1000 0000 0101] sequence to be sent to the transmitter. The key for low energy consumption lies in the fact that in many applications, transmitters are idle most of the time. Therefore, the proposed concept of piezoelectric push-button power generation fits snugly since transmission will only be required when the push button is being depressed, like the case of typical wireless remote controllers. In order to power the RF transmission operation, which takes approximately tens of milliseconds for a complete 12-bit information transmission as shown in Figure 4.6, the electrical energy has to be stored properly in order to ensure complete transmission operation even when the external power source is temporarily unavailable. During this time, the transmission operation is constantly consuming energy until the process is completed where only a minimal amount of energy is needed by the transmitter circuit during the standby mode.

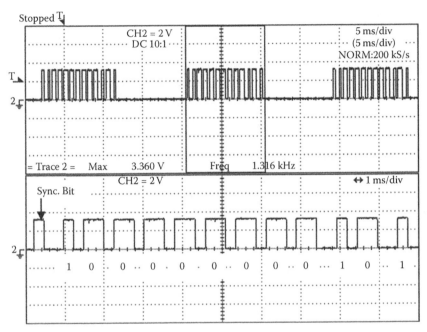

FIGURE 4.6

An enlarged diagram of the 12-bit data sent out by the HT12-E encoder.

4.1.2 Energy Storage and the Power Processing Unit

The PPU converts the AC output voltage from the piezoelectric generator into a DC source through AC-DC full-wave diode bridge rectification. After this, the capacitor connected across the output of the full-wave bridge rectifier doubles as a storage device since energy from the piezoelectric generator is finite and occurs for only a short pulse as well as a filter, which helps to smooth the voltage ripples in the rectified DC voltage. Hence, by choosing a capacitor with a higher capacitance value, the rectified DC voltage can be stepped down to a lower level effectively by the capacitor; at the same time, the capacitor can better smooth the DC voltage and also provide more energy storage for the harvested energy. Thereafter, the capacitor output voltage, the unregulated DC voltage, is further regulated by a voltage regulator circuit into regulated DC voltage.

Due to several design constraints imposed on the impact-based VEH system like miniature size, simple design, and low-cost requirements, a simple voltage regulator is used. According to Dewan et al. [112], if the voltage difference between the input and output of the linear voltage regulator is kept to a minimum, then the linear regulator's efficiency is maximized. This is because a small amount of voltage is dropped across the voltage regulator; hence, little energy is wasted. A switched-mode voltage regulator, on the other hand, suffers from ripples in the output voltage due to its switching rates and has a very high quiescent current compared to a linear voltage regulator, especially in low-power operations where load currents are very low, on the order of microamperes. Although a switch-mode voltage regulator is relatively more complex, it has some superior advantages over a linear regulator, like high power conversion efficiency, voltage step-up and step-down capability, and more. After considering both the advantages and the disadvantages of applying the two types of voltage regulators for research, a linear regulator is more suitable for this VEH research work.

The schematic diagram of the proposed piezoelectric energy harvesting system is shown in Figure 4.7. Referring to Figure 4.7, the principle of operation of the piezoelectric energy harvester for powering a self-powered wireless transmitter is illustrated. When the piezoelectric push button is depressed, around 5 to 7 kV of AC voltage is generated. The generated AC voltage is rectified into DC voltage by the full-wave rectifying diode bridge, and the DC voltage is stored temporarily in the 2.2-μF capacitor. The presence of the capacitor is not only to store harvested energy but also to help clamp down the DC voltage to about 8 V and to smooth the DC voltage to constant voltage. The unregulated DC voltage from the capacitor is regulated by the MAX666 linear regulator. The regulated output DC voltage of MAX666 is 3.3 V, which is then used to operate the HT-12E encoder to transmit 12-bit address/data information via the radio transmitter. The whole piezoelectric push-button transmitter system is mounted on a printed circuit board (PCB) with a total area size of 25 cm^2, and its assembled prototype [113] is shown in Figure 4.8.

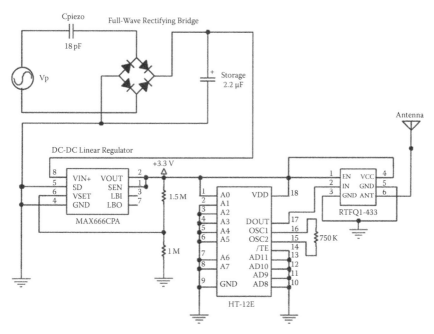

FIGURE 4.7
A circuit schematic of a piezoelectric energy harvesting circuit.

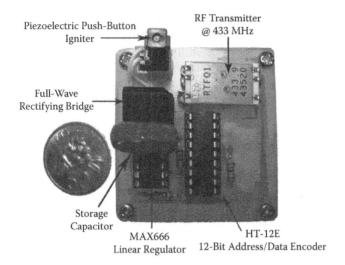

FIGURE 4.8
Photograph of the assembled prototype showing the key components.

Key components of the system are listed in Figure 4.8. They include the piezoelectric push-button igniter, full-wave rectifying diode bridge, 2.2-μF storage capacitor, MAX666 linear regulator, HT-12E 12-bit address/data encoder, and a RF transmitter operating at 433 MHz. A Singapore 50-cent coin has been included for relative size comparison. Comparing the sizes between the piezoelectric push-button transmitter system and the coin as illustrated in Figure 4.8, the requirement of miniature design for the system has been achieved. In addition, the whole assembly is very light, weighing about 12 g. Taking into account all the components used in the circuit design, the total cost of assembly was less than SGD$35.00, which meets the low manufacturing cost requirement. Hence, the small size, light weight, and low-cost features of the piezoelectric push-button transmitter system make it suitable for many wireless applications like a wireless remote controller.

4.1.3 Experimental Results

Several experimental tests were conducted to evaluate the performance of the designed piezoelectric push-button igniter system. The open-circuit AC voltage generated by the piezoelectric igniter as seen in Figure 4.9 is fed into a full-wave diode bridge rectifier, where the AC voltage is converted into DC voltage, and then the DC voltage is stored in the capacitor. The DC voltage waveform at the output of the diode bridge rectifier is shown in Figure 4.10. Referring to Figure 4.10, it can be seen that the peak DC voltage is around 600 V. The zoomed view of the output DC voltage of the full-wave diode bridge rectifier is shown in the bottom waveform of Figure 4.10.

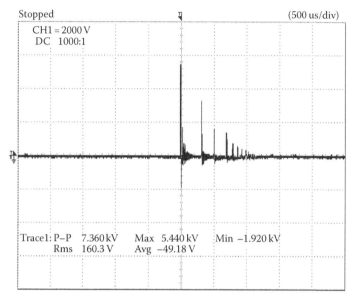

FIGURE 4.9
An open circuit voltage of piezoelectric push-button igniter.

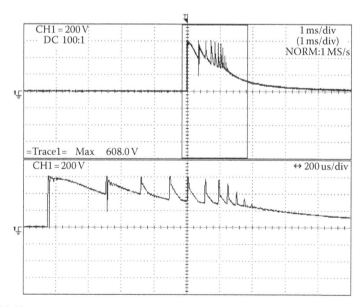

FIGURE 4.10
Output DC voltage of the full-wave diode bridge rectifier.

The sawtooth waveform observed in Figure 4.10 is the rectification of the AC voltage pulses of the piezoelectric push-button igniter. After the AC-DC rectification, the unregulated DC voltage is filtered by a parallel connected capacitor of 2.2 μF added to provide some smoothing effects on the DC voltage as well as to store the harvested energy. The waveforms of the capacitor and the output of the linear regulator are shown in Figure 4.11 where Channel 1 is the DC voltage across the capacitor and Channel 2 is the regulated voltage output from the linear regulator voltage.

It can be seen from Figure 4.11 that the maximum voltage accumulated across the capacitor is 7.84 V, and the capacitor is discharged completely in 300 ms. Using the formula

$$E_{cap} = \frac{1}{2}C V_{cap}^2 \tag{4.8}$$

the amount of electrical energy that is harvested after the piezoelectric push button is depressed is 67.61 μJ. As for the linear regulator, the output DC voltage is regulated at 3.3 V, and it remains constant for around 50 ms before it starts to drop to zero with respect to the capacitor discharge rate. Hence, the challenge is that within the 50 ms of regulated DC 3.3 V, the 12-bit address/data information has to be transmitted from the HT-12E encoder via the radio transmitter. As mentioned, the harvested energy occurs only for a short pulse through actuation of the piezoelectric pulse generator; therefore, in evaluating the performance of our energy harvesting circuit design, the voltage waveforms of the discharging capacitor together with the DC linear

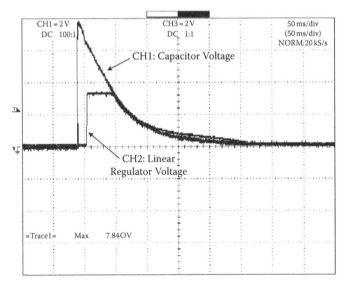

FIGURE 4.11
Voltage waveforms of a capacitor and output of a linear regulator.

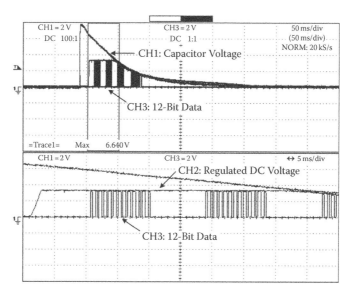

FIGURE 4.12

Output waveforms of a capacitor discharge, 3.3-V DC regulation, and 12-bit address/data transmission.

regulator and the 12-bit address/data information transmitted need to be examined closely.

The results obtained from the energy harvesting circuit design are illustrated in Figure 4.12. Channel 1 in the figure represents the discharging waveform of the storage capacitor. Channel 2 shows the 3.3-V DC signal output waveform from the linear regulator, and Channel 3 represents the 12-bit address/data information transmitted by the HT-12E encoder via the radio transmitter. The energy consumption of the RF unit is calculated based on the operating and standby modes of the RF unit. The total latency time required for 1 digital word transmission is 20 ms; the time during which transmission is active is 10 ms, and the time during which transmission is on standby is 10 ms. During RF unit operating mode and standby mode, the currents drawn by the transmitter are 8 mA and 0.1 μA, respectively. Therefore, the peak powers required during transmission and standby are 2.64 mW and 0.33 μW, respectively. The total maximum energy required for 1 digital word transmission is then 26.4 μJ. Knowing that the harvested electrical energy is 67.61 μJ and the energy required to transmit 1 digital word is 26.4 μJ, it can be calculated that around 2.5 digital words can be transmitted with one push of the piezoelectric push-button igniter. This is verified in Figure 4.12. As shown in the figure, the regulated DC voltage from the linear regulator is sufficient to make at least two complete 12-bit information transmissions at 3.3-V DC. It can therefore be seen that the piezoelectric push-button energy harvesting circuit is able to harness enough energy for two complete transmissions at a constant 3.3-V DC.

4.1.4 Summary

In this impact-based VEH research, special interest is placed on using a piezoelectric push-button igniter as the energy harvester because it is easy and simple to harvest mechanical force energy from human beings. An energy harvesting circuit design for the piezoelectric push-button generator was proposed and implemented efficiently for a wireless RF transmitter. This self-powered wireless transmitter is capable of transmitting a 12-bit digital word information using the mechanical force energy of 15 N harvested from depressing the push button attached to the energy harvesting circuit. Experimental results showed that when the piezoelectric push button is depressed, 67.61 µJ of electrical energy is harvested, and it is sufficient to transmit information for at least 2 complete 12-bit digital words via the RF transmitter unit, which consumes 26.4 µJ of energy for 1 digital word transmission. As such, this research work has successfully demonstrated the feasibility of a completely self-autonomous piezoelectric push-button wireless RF transmitter, which optimizes effectiveness in size, weight, and cost.

4.2 Impact-Based VEH Using Prestressed Piezoelectric Diaphragm Material

Cabling has always been a hassle for applications like the design of a house's lighting system with the need to draw cables from lamps and bulbs to the switches mounted on the walls of the house. The need for in-wall cabling will often result in high costs for the homeowner. Undesirable recabling implications may also arise over time should the cable become faulty. Other than that, the conventional method to power onboard wireless communication and electronic circuitries of the controllers are normally alkaline or rechargeable batteries. One major drawback with these commercially available battery-operated remote controllers is that batteries have a limited energy supply. Every time the remote controllers operate, the energy level of the internal battery source depletes, until some time later when the controllers are no longer able to function. As such, regular maintenance of the controllers is required in order to not create any disruption to the user. To address this issue, an impact-based energy harvesting system using a piezoelectric igniter has been carried out as discussed in Section 4.1. It is a simple and viable solution to harvest mechanical force energy from human beings to sustain the operation of the remote controller. Whenever someone depresses the push-button mechanism, a high-impact force coming from the internal hammer strikes onto the piezoelectric material, which consists of a stack of piezoelectric ceramic layers, and some electrical energy is generated. However, under this high-stress cycling, it is mentioned by Kim et al. [114] that the piezoelectric material stack of the push-button igniter may develop interfacial cracking or buckling, thus shortening its lifetime. In addition, the piezoelectric igniter

FIGURE 4.13
A basic set of a LightningSwitch with one transmitter (left) and one receiver (right).

suffers from two more drawbacks: (1) high input force is required to release the hammer of the push-button mechanism and (2) high output voltage and low output current of a few kilovolts and microamperes are generated due to the piezoelectric stack structure. Because of that, the design and implementation of the power management circuit become complicated.

To overcome these drawbacks associated with the piezoelectric igniter system, another type of impact-based 31-mode piezoelectric generator, operating in mechanical nonresonance, has been explored to harvest impact or impulse forces from human pressing. There are several types of impact-based piezoelectric generators described in the literature [104–106], and there are also some companies like LightningSwitch [115] that apply this piezoelectricity technology on commercial products like a wireless control switch. The self-powered wireless control switch with energy harvesting capability has two significant advantages: (1) flexibility in positioning the power electronic devices at any part of the deployment area that is within the RF communication zone and (2) elimination of the high manpower and material costs in laying wiring cables between the switches and the devices. A basic set of LightningSwitch, consisting of one transmitter and one receiver, is shown in Figure 4.13. Inside the transmitter (see Figure 4.14), LightningSwitch employs

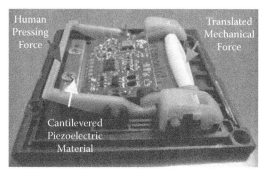

FIGURE 4.14
Internal design of a LightningSwitch transmitter.

a uniquely assembled cantilever design to mount and excite the commercially available piezoelectric transducer material for harvesting vibrational energy generated from human pressing.

Referring to Figure 4.14, whenever a user depresses the commercially available switch equipped with the energy harvesting feature, the harvested impact force is not directly applied to the energy harvesting material (i.e., piezoelectric PZT material), instead it is transferred from the human depressing point through an in-built mechanical cantilevered structure to indirectly activate the material. The custom design mechanical structure translates the applied mechanical force onto the edge of the cantilevered piezoelectric transducer material. As the mechanical force is applied onto the piezoelectric material, it bends until the mechanical structure releases the piezoelectric material; thus, mechanical oscillating vibration is generated. The vibration energy is then converted into electrical energy using the cantilevered piezoelectric material. In the LightningSwitch design, the whole energy conversion process involves many different stages from human depressing force to harvested electrical energy. During these conversion stages, a certain portion of the harvested mechanical energy is lost due to an energy efficiency drop at each of the individual stages. In addition, the uniquely assembled cantilever design used in LightningSwitch products, which employs the extra complicated mechanical structure design, would also incur more cost to develop and manufacture it.

Unlike the commercially available products and research prototypes, which employ a complicated energy harvesting mechanism to excite its 31-mode piezoelectric generator, a relatively new concept of energy harvesting from depressing a prestressed piezoelectric diaphragm material has been proposed in this research to generate electrical power for a wireless RF transmitter and its power management circuit to switch on/off electrical appliances such as lighting, fans, and so on, in a wireless manner. The main objective of this research is to fulfill a self-powered switch with a less-complicated as well as less-costly energy harvesting technique. This is achieved by removing the excessive components of the energy harvesting mechanism.

4.2.1 Description of Prestressed Piezoelectric Diaphragm Material

The impact-based energy harvesting mechanism used in this research is simply a prestressed piezoelectric diaphragm material manufactured by Face® International Corporation based on the THUNDER (THin layer UNimorph ferroelectric DrivER and sensor) technology originally developed by NASA in conjunction with the RAINBOW (reduced and internally biased oxide wafer) design effort [116]. The prestressed piezoelectric diaphragm material, as shown in Figure 4.15, is electrically poled in the 31 coupling mode, and it is initially curved, arc shaped, and rectangular, which elongates when a force is applied to the top of the arc. The elongation causes strain in the active material, which produces a voltage. The device is simply supported and allows for movement only in the lateral direction.

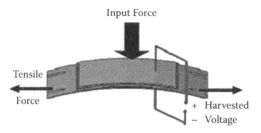

FIGURE 4.15
A diagram of the prestressed piezoelectric diaphragm material, TH7R.

As an input force is applied at the centre of the warped structure dome-shaped piezoelectric strip, a resultant tensile force is generated in the piezoelectric material along the 1-axis and voltage is harvested and poled along the 3-axis, as illustrated in Figure 4.15. Unlike the piezoelectric igniter that operates in the parallel compression mode or 33-mode of operation, this transverse configuration maximizes the piezoelectric transducer efficiency in converting mechanical energy to electrical energy in a low-force environment. The innovative part of this research is the use of the natural characteristic and construction of the prestressed diaphragm material as an energy harvesting material (its unique characteristic is that when the material is depressed, it is deformed, and it generates electrical energy) as well as a switch (its prestressed construction creates a bouncing reaction whenever the material is depressed at the centre).

Referring to Figure 4.16, the THUNDER piezoelectric device consists of a piezoelectric ceramic layer, PZT, and a backing metal layer, held together with a polyimide adhesive as described by Bryant [117]. Several researchers have investigated this type of device, and literature is widely available on its actuation and manufacturing details. However, very few researchers have examined its potential as an energy conversion mechanism [118]. As an energy harvesting device, few publications are available: Ramsay et al. [54] demonstrated the feasibility of utilizing the power generated in a bio-MEMS (bio-microelectromechanical system) application; Mossi et al. [119] compared different configurations of the device for actuation and energy harvesting; and Shenck et al. [46] utilized this device for harvesting energy by mounting

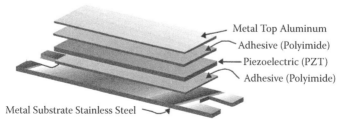

FIGURE 4.16
Construction of a THUNDER TH7R

TABLE 4.1

Technical Specifications of TH7R

TH7R Dimensions and Physical Properties		
Weight	g	18
Dimensions	mm	95.3 × 73.4 × 0.53
PZT thickness	mm	0.25
Static capacitance	nF	166
Maximum voltage	V	300
Vertical displacement	mm	9.55

it on a shoe. Danak et al. [120] also researched ways to optimize the design of an initially curved PZT unimorph power harvester. A mathematical model was created that predicts the power output of the device. From this model, relationships between generated charge and initial dome height, substrate thickness, PZT thickness, and substrate stiffness were established.

THUNDER transducers are commercially available in a variety of sizes and force displacement characteristics, each with very different electromechanical characteristics. According to Danak et al. [120], the power available from a flat piezoelectric transducer under 31-mode bending excitation is generally proportional to the volume of the material and the vertical displacement induced. Therefore, while constrained by the size, comfort, and vertical displacement experienced by the human pressing, the volume of PZT piezoceramic is maximized when selecting the appropriate transducer. Hence, the TH7R piezoelectric device is chosen, and its technical specifications are shown in Table 4.1.

4.2.2 Characteristics and Performance of THUNDER Lead-Zirconate-Titanate Unimorph

The unique prestressed characteristic and convex shape of the THUNDER piezoelectric material is used as a natural "press-and-release" mechanical switch. The natural "press-and-release" process of the piezoelectric material illustrated in Figure 4.17 consists of displacing the transducer from its equilibrium position, reaching maximum displacement (stress), before allowing

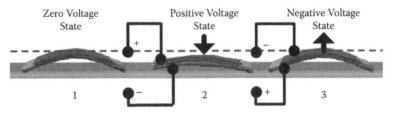

FIGURE 4.17
A depressing cycle of the prestressed piezoelectric diaphragm material.

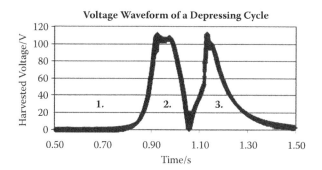

FIGURE 4.18
Voltage waveform of a depressing cycle of the prestressed piezoelectric diaphragm material.

the transducer to return to its equilibrium position via the opposite direction. The change in direction of displacement, and hence direction of stress, causes a reverse in polarity, and an AC voltage is generated at the output of the piezoelectric material. The impact-based energy harvesting process seen in Figure 4.17 can be described as follows: Initially, at zero voltage state labelled as 1, the piezoelectric material is not depressed, so it can be seen from Figure 4.18 that no voltage is generated by the piezoelectric material. On depressing the material, which is illustrated by the positive voltage state labelled 2 in Figure 4.17, the material is flattened, and the electrical voltage developed across the material can be read from Figure 4.18 to be 110 V. Energy accumulated in the piezoelectric material is available for harvesting by the designed power management circuit and the connected wireless transmitter load. Similarly, on releasing the material as illustrated by the negative voltage state labelled 3 in Figure 4.17, the material bounces back to its original prestressed state, a rectified electrical voltage of 110 V is developed across the material, and energy is stored in the material for harvesting.

The natural bouncing characteristic and power generation ability of the material has been utilized by this research to simplify the structure of the energy harvesting mechanism. By doing so, it is no longer necessary to have any additional mechanical structure, like the LightningSwitch case shown in Figure 4.14, to translate the human pressing force into electrical energy. This is a great advantage of this research over the commercial products available on the market. Without the need of an extra energy translation mechanism, this proposed impact-based energy harvesting mechanism using a prestressed piezoelectric diaphragm material reduces the cost of manufacturing the switch to lower than that of the more complicated LightningSwitch design. In summary, this research work, which uses a batteryless wireless control switch and prestressed piezoelectric diaphragm material, is less complicated to implement, is less susceptible to wear and tear due to less mechanical parts, and enables more cost savings than similar products available in market.

To determine the characteristic of the prestressed piezoelectric diaphragm material, different input forces—hard (~10 N), normal (~5 N), and light

(~1 N)—provided by a human's thumb pressing are applied onto the piezo-electric material. For each applied input force, the output AC voltage V_{ac} of the piezoelectric material is rectified by a diode bridge into DC voltage V_{dc} for various loading resistances R_{load} to calculate the equivalent electrical energy being harvested. In this case, the voltage drop across the diode bridge is very low as compared to V_{ac}; hence, it is reasonable to neglect the voltage loss in the diodes. Since the voltage generated by the piezoelectric material is time-variant, as can be seen in Figure 4.18, the electrical energy harvested $E_{harvested}$ has to be calculated based on the following equation:

$$E_{harvested} = \int \frac{|V(t)|^2}{R_{load}} dt \qquad (4.9)$$

where $V(t)$ is the instantaneous voltage taken from the generated voltage waveform. In order to solve the integral, the oscilloscope readings are extracted into an ASCII file and analyzed using Microsoft Excel. Figure 4.19 demonstrates the method of estimating $E_{harvested}$ based on the integral of $V(t)$ by using a first-order approximation.

Referring to Figure 4.19, since the oscilloscope reads in discrete measurements, the integral of $V(t)$ is estimated by approximating a straight line joining all the measured points, resulting in many trapeziums of equal width, and summing the areas of each trapezium. Hence, the integral of $V(t)$ can be calculated as follows:

$$\int |V(t)| \, dt = \sum \frac{1}{2} (t_{x+1} - t_x) [V(t_{x+1}) + V(t_x)] \qquad (4.10)$$

The accuracy of the estimation depends on the sampling frequency. On average, the sampling frequency of the oscilloscope is 5 kHz; hence, each interval is 0.2 ms, which is sufficiently accurate for approximating the integral of $V(t)$ and therefore the electrical energy harvested $E_{harvested}$. Based on Equations 4.9 and 4.10, the maximum harvested energy from the piezoelectric material with

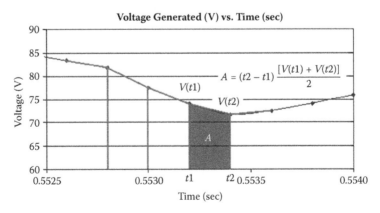

FIGURE 4.19
Estimating integral using a first-order approximation.

TABLE 4.2

Maximum Energy Available for Harvesting under Various Input Forces

Input Force (N)	Peak Voltage (V)	Time Span (s)	Energy Harvested mJ
Light / 1	73.3	1.112	0.88
Normal / 5	86.7	0.846	1.23
Heavy / 10	112.5	0.721	2.08

internal capacitance C_{piezo} of 164 nF is tabulated in Table 4.2 for various input forces provided by a human's thumb pressing.

Apart from determining the optimal load condition, the optimal capacitance that maximum energy can be transferred from the piezoelectric material and stored into the external storage capacitor has to be investigated as well. In this experimental test, the load connected to the output terminal of piezoelectric material is a capacitor with capacitance values ranging from 150 nF to 33 μF. The experiments are conducted to investigate the performance of the piezoelectric material in terms of its generated peak voltage and harvested energy for different capacitor values, and their experimental results are illustrated in Figures 4.20 and 4.21.

Referring to Figure 4.20, it can be observed that the peak voltage generated across the capacitor falls with the capacitance value of the external capacitor. Based on the principle of conservation of charges, $Q = CV$, as the capacitance value increases, while the generated charges remain unchanged, the voltage developed across the capacitor decreases. This is the voltage clamping effect of the capacitor. Similarly, with reference to Figure 4.21, it can be seen that the electrical energy stored in the capacitor decreases as the capacitance value increases. According to Shenck [101], if there is impedance mismatch between the source capacitance of piezoelectric material and the load capacitance, energy loss is bound to occur in the energy transfer process. Hence, transferring energy from a fixed capacitor (source capacitor) to an increasingly large capacitor will result in greater mismatch and higher losses. Among the

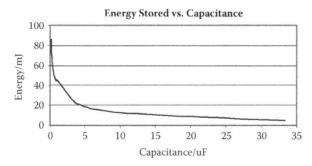

FIGURE 4.20
Peak output voltage generated at various capacitance values.

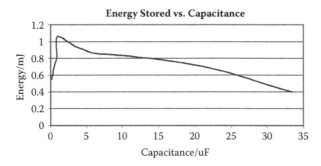

FIGURE 4.21
Harvested energy for various capacitance values.

various capacitance values, 3.3 μF yields the highest stored energy of 1.1 mJ and so does its generated voltage of around 85 V. This high output voltage might be a great challenge to the design of the power management circuit. As such, it is necessary to make a compromise between the harvested energy and the generated voltage to meet the energy requirement of the wireless load.

4.2.3 Power Management Circuit

In order to determine the technical specifications of the power management circuit, the power requirements of the RF transmitter have to be determined. The transmitter power requirement is more important as it is the load to which the piezoelectric transducer supplies. Since the input voltage and current vary according to the transmission range and data rate, there is a need to determine the relationship between the input voltage and transmission range, as well as the corresponding input current. An experiment was conducted to investigate the relationship as shown in Table 4.3.

TABLE 4.3

Power Consumption of RF Transmitter Load

V_{ee} (V)	I_{cc} (mA)	Minimum Transmission Time (ms)	Range (m)	Power (mW)	Energy (mJ)
1.5	0.280	105	0.0	0.42	0.044
1.6	0.365	95	6.5	0.58	0.055
2.0	0.450	84	—	0.90	0.076
2.4	0.580	73	10.0	1.39	0.102
2.8	0.710	60	—	1.99	0.119
3.6	0.975	57	15.5	3.51	0.200
4.0	1.120	55	—	4.48	0.246
4.4	1.210	53	18.5	5.32	0.282
5.0	1.410	50	20.0	7.05	0.353
6.0	1.740	48	—	10.44	0.501
7.0	2.060	45	24.0	14.42	0.649

Since the wireless control switch is designed for indoor conditions, the communication range of the transmitter is set to be around 20 m, hence the power consumption of the load read from Table 4.3 is 5 V and 1.41 mA, and the time duration to transmit three 13-bit signals is 50 ms. Apart from obtaining the power requirement of the RF transmitter load, the duration for a successful transmission is also required to determine the total energy required. Thus, the energy stored in the capacitor must be around 0.35 mJ to power the load at more than 5 V for at least 55 ms. From these specifications, with reference to Figures 4.20 and 4.21, the 10-μF capacitor is the most suitable value as it has a peak voltage of 12 V and is able to store 0.83 mJ of energy when fully charged. The principle of operation of the proposed energy harvesting system is described as follows: Whenever the switch is depressed, the mechanical force is applied onto the prestressed piezoelectric material, and some of the mechanical energy is converted into electrical energy. The harvested electrical energy from one single pressing is stored in an energy storage device until a certain preset threshold level is reached before the stored energy is released through a power management circuit to power up the RF transmitter. The RF receiver, which is connected to the electrical appliance, receives the transmitted signal and switches the appliance on/off by controlling its power supply. The schematic drawing of the proposed impact-based energy harvesting system and its power management circuit is shown in Figure 4.22.

Referring to Figure 4.22, the operation of the circuit can be explained. Charges generated by the piezoelectric transducer are first transferred into capacitor C2, while the regulator and transmitter (load) are isolated by the N-MOSFET (N-metal-oxide-semiconductor field-effect transistor) N-MOS Q2, which cuts off the load ground from the source ground. The Zener diode Z1 disconnects the P-MOSFET (P-MOS) Q1 until the voltage across capacitor C1 exceeds its reverse breakdown voltage plus the threshold voltage. Once Q1 is turned on, the voltage across R2, adjustable by the potential divider formed by R1 and R2, exceeds the threshold voltage of Q2 and turns on N-MOS Q2. Thus, the source ground and load ground are connected, and C2 starts discharging to the regulator and the RF transmitter. R3 acts as the latch to ensure

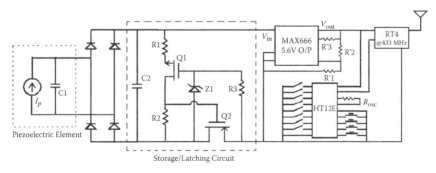

FIGURE 4.22

Schematic drawing of the proposed impact-based energy harvesting system.

that P-MOS Q1, and in turn N-MOS Q2, remain on when the voltage across C2 drops below the Zener diode's breakdown voltage. This is because once the source and load ground are connected, current flows through R3, thus maintaining a voltage at the gate of P-MOS Q1 to latch the regulator to the capacitor. The capacitor C2 stops discharging once the voltage across it falls to an extent that the voltage across R2 is below the threshold voltage of Q2. The duration of the discharge must be at least as long as the time required to transmit three successive RF signals, so that the RF receiver can acknowledge the transmission and turn on/off the electronic appliance to which it is attached.

4.2.4 Experimental Results

To demonstrate the feasibility of this proposed impact-based energy harvesting system using a prestressed piezoelectric diaphragm material, the hardware prototype of the batteryless wireless control switch has been implemented [121] as shown in Figure 4.23.

The component values of the schematic diagram of the proposed impact-based energy harvesting system shown in Figure 4.22 are determined with the following considerations: The first component to decide is capacitor C2. As analyzed previously, 10 μF was chosen as it has a peak voltage of about 12 V; it also has a suitable discharge time constant and stores sufficient energy for wireless transmission ($\sim$0.836 mJ). A complementary MOSFET (N-MOS and P-MOS) is chosen as it has a low threshold voltage (1.0 V) and is conveniently placed together as a single component. Therefore, a Zener diode with a reverse

FIGURE 4.23
Prototype of a proposed batteryless wireless control switch using prestressed piezoelectric diaphragm material.

breakdown voltage of 11.0 V (12 − 1 V) is selected. This ensures that the capacitor is charged to at least 12 V before it discharges to the regulator. However, a high breakdown voltage also implies that if the transducer is compressed too lightly and the maximum voltage of the capacitor fails to reach 12 V, no power will be supplied to the transmitter. R1, R2, and R3 should be as large as possible so that minimum current flows through them, and resistive losses are minimized. As mentioned, $R2/(R1+R2) > V_{GS}/V_{reg}$, and with V_{reg} at 5.0 V, $R1 = 10$ MΩ, $R2 = 5$ MΩ, and $R3 = 10$ MΩ.

Experimental results shown in Figure 4.24 illustrate the outcome of the designed power management circuit. The voltage waveforms in Figure 4.24 are measured across the storage capacitor and the voltage regulator of the power management circuit. During one depression cycle: (1) As the human's finger presses down onto the material, the harvested energy is stored in the capacitor and not released to the load as the source and load are disconnected. The capacitor's voltage rises to around 7 V. (2) As the human's finger stops pressing, energy is harvested again and stored in the capacitor. The capacitor's voltage continues to rise to 13 V, which is more than the preset voltage level, and energy is released to the load. Once the source and load are connected, energy is supplied to the load, as can be seen in Figure 4.24 where the output voltage of the load has been regulated by the voltage regulator.

The circuit design of the receiver unit is shown in Figure 4.25. The design basically is comprised of an RF receiver circuit, a timer, and a JK flip-flop. The JK flip-flop is used to implement the simplest toggle design by connecting its inputs together. A 555 timer is used as a timing device. The decoder has a signal VT, which only turns high for a short duration when a signal from

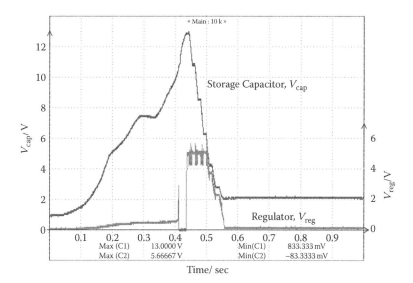

FIGURE 4.24
Voltage waveforms across a storage capacitor and voltage regulator.

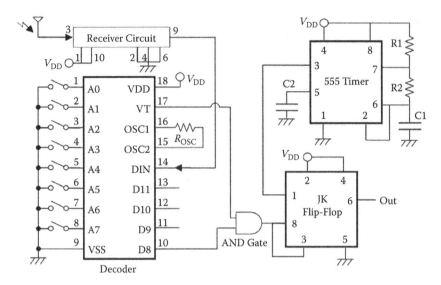

FIGURE 4.25
A schematic circuit diagram of the RF receiver circuit.

the transmitter is received. It is basically a square pulse with a width of about 0.11 s. Thus, to ensure that the JK flip-flop is triggered during the rising edge of the clock when the signal VT sends in a pulse, the frequency of the timer is set to 9 Hz. For the clock cycle waveform and the square pulse signal VT observed in Figure 4.26, the duty cycle is set to 0.5 so that the rising edge is the same for each cycle, and the timer components are related to the frequency, $f = 1/0.11$,

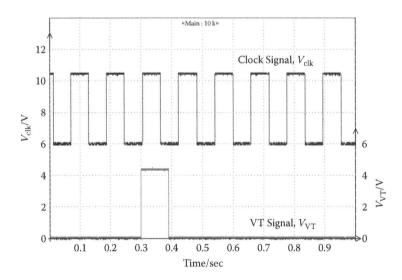

FIGURE 4.26
Voltage waveforms of VT and the clock cycle generated by a 555 timer.

by the equation: $(R1 + 2R2) = 1.44/(f * C1)$. The components chosen are $R1 = 1$ KΩ, $R2 = 80$ KΩ, $C1 = 1.0$ μF, $C2 = 10$ nF. Oscillator resistance (Rosc) is fixed at 82 KΩ because the RF transmitter is operating at 433 MHz. The output of the JK flip-flop can be connected to a relay to turn the connected electrical appliances on/off.

Referring to Figure 4.25, it can be seen that an AND gate is added to D8 of the decoder. This means that if the transmitter sends a "high" on the corresponding data bit, it toggles this circuit. Thus, this circuit is considered to be "tagged" with the first data bit and by the same logic, four more receivers each tagged to the other four respective data bits can be built. By doing so, a single transmitter will be able to control up to four different electronic devices. In addition, this design saves the cost of having to buy one switch for every appliance, especially when it comes to appliances that are usually grouped together, such as a set of four lights in a single room. This will also eliminate the hassle of having to find the individual switch for a particular appliance, as well as having to allocate space for four switches.

4.2.5 Summary

In this impact-based energy harvesting research, a batteryless wireless control switch using a prestressed piezoelectric diaphragm material has been proposed for applications in warehouses, commercial buildings, and so on, to cut high wiring costs, including material, and labour expenses in laying power and signal cables from control switches all the way to the electrical appliances can be reduced tremendously. In addition, no maintenance cost and effort are required for the wireless control switches. The self-powered wireless control switch is also highly suitable for those appliances located in remote locations that are too difficult or hazardous to access and also suitable for applications where multiple electrical appliances have to be controlled simultaneously. Unlike the piezoelectric igniter system, the prestressed piezoelectric diaphragm material is more applicable for low-force environment, and its output voltage is relatively low. Since the electrical power throughput of the piezoelectric material is less than the power requirement, an energy storage and supply circuit has been implemented. By doing so, 0.83 mJ of energy is first stored in the capacitor until the preset voltage level of 12 V, and the stored energy is supplied to the RF transmitter load through a voltage regulator of 64% efficiency. From the experimental test results obtained, the harvested energy output from the voltage regulator of 0.53 mJ is sufficient to power one successful wireless transmission for a distance of 20 m.

5

Hybrid Energy Harvesting System

Small-scale energy harvesting (EH) is a fast-growing research solution for powering a wireless sensor node [122–125]. However, EH itself has an inherent problem, which is the intermittent nature of the ambient energy source. It is possible that the operational reliability of the sensor node becomes compromised for prolonged unavailability of the ambient energy source. To augment the reliability of the sensor node, hybrid energy harvesting (HEH) is proposed in this chapter. For the HEH approach, a second ambient energy source, which is available in the same environment as the first energy source, is harvested to supplement the energy supply of the wireless sensor node. By doing so, more energy can be potentially harvested, such that the wireless sensor node is able to perform more power-intensive operations, such as increasing its transmission rate or increased support for external sensors and peripherals.

The concept of HEH has been recently discussed in the literature [35, 126–129] as a potential micropower supply solution to minimize the size of the energy supply as well as to extend the operational lifetime of the wireless sensor node. Researchers have considered a number of methods to combine different small-scale EH sources, and these methods can be classified into four main categories.

- *Type 1: HEH using two different EH mechanisms on the same platform.* Tadesse et al. [126] and Khaligh et al. [127] present a mechanical combinatory EH device structure of two different EH mechanisms on the same platform to harvest from the same vibration energy source. Due to the drastic difference in matching the internal impedances of the electromagnetic (Ω) and piezoelectric (MΩ) EH mechanisms with the external load resistance, it presents difficulty in combining the output power from the two different mechanisms; hence, different power converters are required to process the harvested power separately.

- *Type 2: HEH using an electronic switch/multiplexer to switch between EH sources.* Guilar et al. [128] and Lhermet et al. [129] proposed to combine EH using an electronic switch/multiplexer to switch between the two energy sources. Whenever both energy sources are present simultaneously, based on the priority given by the power management circuit, typically the higher-power source, only one of the two energy sources would then be harvested. Hence, it is not possible to harvest energy from both energy sources simultaneously.

- *Type 3: HEH using an individual power converter for each EH source.* According to Park et al. [35], each energy source has its own power management circuit; hence, it allows energy to be harvested simultaneously from both energy sources. However, as the number of converters increases, this scheme becomes more complex and bulky with higher component counts, and the power loss in the individual converter control circuitry becomes especially significant for micropower EH sources.

- *Type 4: HEH by directly connecting the energy sources in parallel/series configuration.* Energy sources are directly connected together, and they share a common power management circuit. Compared with the energy sources selection method used in Type 2, this scheme allows energy to be harvested simultaneously from both energy sources. In addition, only one power electronic-based converter with a simple and low-power control circuitry is required instead of dedicating each individual energy source with a power converter described in Type 3. However, the challenge of this approach is that there could be an impedance mismatch issue among the integrated energy sources.

Of the four HEH methods mentioned, Types 1 and 2 are not adopted for the HEH research of this chapter. The reason is that the Type 1 approach does not have the advantage of harvesting from another energy source. As for the Type 2 approach, even though there are two different energy sources available for harvesting, the HEH system does not have the capability to harvest from both the energy sources at one time. Hence, in this chapter, two types of small-scale HEH schemes are investigated: (1) hybrid of wind energy harvesting (WEH) and solar energy harvesting (SEH) schemes for outdoor application and (2) hybrid of indoor ambient light and thermal energy harvesting (TEH) schemes for indoor application.

The first HEH system, illustrated in Section 5.2, is designed according to the Type 3 approach of combining the two energy sources. In this Type 3 approach, each EH source is required to have its own unique power management unit to condition the power flow from the energy source to its output load. However, this is not the case for the fourth HEH method, which is utilized by the latter HEH system and illustrated in Section 5.3. The proposed HEH system requires only one power management unit to condition the combined output power harvested from the solar and thermal energy sources. By avoiding the use of different power management units for multiple energy sources, the number of components used in the HEH system are decreased, and the system's form factor, cost, and power losses are thus reduced. Before proceeding into the details of the two HEH systems, the SEH system common to both HEH systems is first investigated.

5.1 Solar Energy Harvesting (SEH) System

Several mathematical models exist in the literature [130–132] that describe the operation of photovoltaic (PV) cells, from simple to more complex models that account for different reverse saturation currents. In this chapter, an electrical circuit with a single diode (single exponential) is considered as the equivalent PV model, which consists of n_s number of PV cells in series, as shown in Figure 5.1.

Assuming that the shunt resistance R_{sh}, as shown in Figure 5.1, is infinite, the current-voltage (I-V) characteristic of the PV module can be described with a single diode as the four-parameter model given by Equation 5.1 [130],

$$I_{pv} = I_L - I_o \left[exp \left(\frac{V_{pv} + I_{pv} R_s}{n_s V_t} \right) - 1 \right]$$

(5.1)

where I_L is the light-generated current (A) and I_o is the dark/reverse saturation current of the p-n diodes (1×10^{-9} A). R_s is the series resistance of the PV module, and V_t is the junction terminal thermal voltage (V) depending on the cell absolute temperature, which is defined as

$$V_t = \frac{k T_c}{q}$$

(5.2)

where T_c is the cell absolute temperature (K), k is the Boltzmann constant (1.3807×10^{-23} J K^{-1}), and q is the charge of the electron (1.6022×10^{-19} C). The ultimate goal is to determine whether the power harvested by the PV module is able to power the wireless sensor node; hence, it is crucial to estimate the electrical power throughput of the PV module by leveraging on the relationship between the current and voltage of the PV module expressed by Equation 5.1. Referring to Equation 5.1, it can be deduced that the voltage drop across the series resistance, $V_{Rs} = I_{pv} R_s$, is comparably much lower than the output PV voltage V_{pv} due to the very low PV current I_{pv} on the order of microamperes flowing through the small series resistance R_s of a few ohms; thus the $I_{pv} R_s$ term in Equation 5.1 can be neglected during the formulation

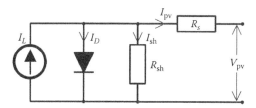

FIGURE 5.1
Equivalent electrical circuit for a photovoltaic module.

of the output power of the solar panel $P_{pv}(V_{pv})$, which is expressed as follows:

$$P_{pv}(V_{pv}) = V_{pv}I_{pv} = V_{pv}I_L - V_{pv}I_o\left[exp\left(\frac{V_{pv}}{n_s V_t}\right) - 1\right]$$

$$\approx V_{pv}I_{sc} - V_{pv}I_o\left[exp\left(\frac{V_{pv}}{n_s k T_c/q}\right)\right] \tag{5.3}$$

Note the term $exp(V_{pv}/n_s V_t) \gg 1$ and the light-generating current, $I_L \approx I_{sc}$ [130]. The harvested PV power $P_{pv}(V_{pv})$, as expressed in Equation 5.3, is formulated as a function of the PV voltage V_{pv}, and it can be estimated based on the technical characteristics of the PV module and the environmental variables such as light irradiance and ambient temperature. Based on Equations 5.1 and 5.3, the current voltage (I-V) and power voltage (P-V) curves of a solar panel at a particular solar irradiance and operating temperature are plotted in Figure 5.2.

Referring to Figure 5.2, there exists a particular operating voltage and current of a PV module under a certain irradiance and temperature that yields the maximum power throughput P_{max}. The maximum power point (MPP) of the PV module seen in Figure 5.2 corresponds to a specific operating voltage V_{mppt} and current I_{mppt}. Various maximum power point tracking (MPPT) techniques have been discussed in the literature [133, 134] to operate the PV module at its MPPs. These MPPT techniques include perturbation and observation (P&O), incremental conductance (IncCond), constant voltage (CV), fractional open-circuit voltage (FOCV), and others.

Among the MPPT techniques, the P&O method is most commonly used by the majority of researchers in large-scale PV systems [74–87]. It is an iterative

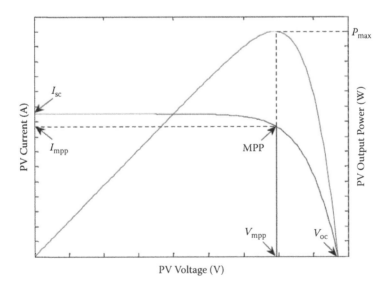

FIGURE 5.2
Maximum power points on I-V and P-V curves of a photovoltaic module.

method of obtaining the MPP. It measures the PV characteristics and then perturbs the operating point of the PV module towards the maximum point when $dP_{pv}/dV_{pv} = 0$ is reached. IncCond is an alternative to the P&O method proposed by Hussein et al. [135]. It is based on differentiating the PV power P_{pv} with respect to voltage V_{pv} and setting the result to zero. The maximum point is reached when the opposite of the instantaneous conductance, $G = I_{pv}/V_{pv}$, is equal to its incremental conductance dI_{pv}/dV_{pv}. According to Esram et al. [133], the P&O and IncCond techniques are the most effective MPPT techniques, harvesting the most energy in comparison. This is because both MPPT techniques have the ability to track the true MPP of the PV module accurately under any solar irradiance level. However, the implementation of these MPPT techniques becomes complex and expensive. They require the use of energy-hungry devices like microcontrollers and some sensory circuitries, that is, voltage and current to compute, process, and track the desired output power in every processing iteration. In addition, at steady state, the operating point of the PV module tends to oscillate around the MPP, thus giving rise to wasting some harvesting energy [74].

In contrast, the CV technique is by far the simplest MPPT technique that can be implemented. The operating voltage of the PV module V_{pv} is kept near the PV's MPP by matching it to a predetermined reference voltage. The reference voltage is chosen to be close to the MPPT voltage V_{mppt}. However, according to Faranda et al. [134], the CV technique is mentioned to be the least-effective MPPT technique. This is because the MPPT voltage tends to shift with the varying solar irradiance and temperature. Hence, there is a small voltage range where MPP occurs for the different operating conditions. Intuitively, the CV technique will only yield an approximate MPP. The FOCV is an alternative to the CV method that is also a simple and cheap solution. It is based on the voltage of the PV module at the MPP, $V_{mppt} = kV_{oc}$, which is approximately linearly proportional k to its open-circuit voltage V_{oc}. During the SEH process, the normal operation of the PV system is interrupted to measure the open-circuit voltage V_{oc} of the PV module. This is done by disconnecting the PV module from the electrical load with an additional electronic switching circuitry operating at a certain frequency. Once the open-circuit voltage is obtained, it is multiplied with a predefined factor k to get the measured MPPT voltage V_{mppt}. The drawback with the FOCV method is that the interrupted system operation yields power losses when scanning the entire control range.

To overcome this drawback, a pilot PV cell was proposed by Brunelli et al. [136]. The pilot cell is supposed to share similar characteristics as the EH PV module to obtain the open-circuit voltage of the PV module through the pilot PV cell. Therefore, it is not necessary to disconnect the PV module from the load in order to obtain the open-circuit voltage. However, the challenge with this approach is that it is difficult to source for a pilot cell that has the exact characteristic as the EH PV module. In short, each MPPT technique has its own pros and cons. Hence, for small-scale EH, it is crucial to choose an appropriate MPPT technique so the energy harvesting system is not overloaded but at the same time achieves the MPPT effect.

5.2 Composite Solar, Wind (S+W) Energy Sources

To understand the functionality and performance of the proposed HEH system in a better way, the circuit architecture of a wind-cum-solar-powered wireless sensor node is presented in Figure 5.3. The HEH wireless sensor node consists of the WEH and SEH subsystems connected in a parallel configuration so that the harvested electrical power from either wind or solar energy source, P_{WEH} or P_{SEH}, respectively, can be used to charge the energy storage device.

A similar HEH research work was discussed by Park et al. [35], Ambi-Max, which harvests simultaneously from two energy sources to sustain the operation of the sensor nodes. Each energy source is allocated to charge its own supercapacitor, and the energy stored in these supercapacitors is then transferred to the lithium ion battery. Although simultaneous charging from two energy sources can be achieved, the drawback is that the physical size of AmbiMax becomes too large, bulky, and heavy as compared to a miniaturized wireless sensor node.

This chapter discusses the use of one energy storage device for simultaneous charging from two energy sources and to develop a miniaturized HEH system for portability and concealment in security applications [138]. Among the commonly available renewable energy sources, solar and wind are the

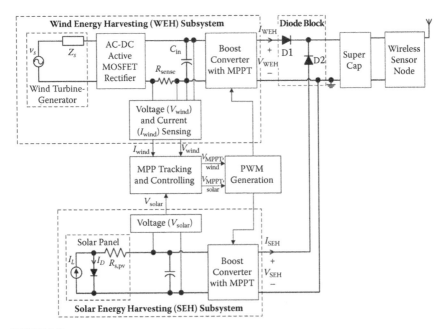

FIGURE 5.3
Functional block diagram of HEH wireless sensor node. (AC-DC, alternating current-direct current; MOSFET, metal-oxide-semiconductor field-effect transistor; PWM, pulse width modulation.)

most abundant energy sources in an outdoor environment, so they have been chosen to be utilized in the proposed HEH system. During the night, when the sun has set, wind energy that is mostly available throughout the whole day is present for harvesting. Therefore, it can be seen that the wind and solar energy sources complement one another. They form an ideal pair for an HEH system in powering wireless sensor nodes deployed in the outdoor environment. However, the electrical characteristics of the wind and solar energy harvesters are different from one another. When these energy harvesters are combined directly, it is bound to cause a problem of internal impedance mismatch between them. This will result in very poor power transfer efficiency between the EH sources and the electrical load.

To overcome this problem, a Type 3 HEH scheme is proposed where each energy source (i.e., wind and solar) has its own power management circuit for performing MPPT. This ensures simultaneous charging of the energy storage device as well as powering of the wireless sensor node. The rest of the section is organized as follows: Section 5.2.1 describes the electrical power generation from the WEH subsystem. Section 5.2.2 details the design of an efficient power management circuit to perform MPPT for the SEH subsystem. Section 5.2.3 illustrates how the WEH and SEH subsystems are interfaced to achieve the proposed HEH system. Following that, the experimental results of the optimized HEH wireless sensor node prototype are depicted in Section 5.2.4, with the summary reported in Section 5.2.5.

5.2.1 Wind Energy Harvesting Subsystem

The proposed HEH wireless sensor node is designed to be deployed in a sample remote sensing area where its environment has four seasons throughout the year (i.e., spring, autumn, summer, and winter). The environmental condition of the deployment area for each season differs from the other seasons. Take, for example, during the summer season, there is lots of sunshine, but only some gentle breezes. Conversely, when it comes to wintertime, the duration of sunlight becomes short, whereas the wind gets stronger. According to the global solar power map and Canada's wind energy atlas (illustrated by Solar Power [139] and the *Canadian Wind Energy Atlas* [140], respectively), it is observed that the average sun hour and average wind speed in the northern part of the world like Toronto (Canada) are 1.0 peak sun hour and 4 m/s, respectively. The term *peak sun hour* represents the average amount of sun available per day throughout the year. The total amount of solar radiation energy can be expressed in hours of full sunlight per square metre (1000 W/m^2) or peak sun hours [139]. Over a time span of 12 hours per day, 1 sun hour works out to be an average solar irradiance level of around 80 W/m^2. With reference to the WEH system designed and developed in Chapter 2, at an average wind speed of 4 m/s, it can be read from Figure 2.4 that the electrical power harvested by the wind turbine generator at its maximal point is around 18 mW.

Based on the power analysis of the WEH system described in Section 2.1.3.2, it can be seen that the WEH system with a MPPT scheme is able to harvest four times more electrical power than its counterpart WEH system using a standard power management circuit. Although the developed WEH system is validated to deliver more electrical power, a larger energy storage device is required to store the greater of harvested energy. The stored electrical energy is used to prolong the operational lifetime of the sensor node in times when wind is not available. In this chapter, other than relying on the energy storage device to sustain the sensor node's operations, another EH system to harvest solar energy in the same environment as the WEH system is explored and then developed for proof of concept. The role of the additional SEH system is to augment the reliability and performance of the wind-powered wireless sensor node.

5.2.2 SEH Subsystem

An overview of the proposed SEH subsystem is illustrated in Figure 5.3. In the SEH subsystem, a small solar panel with a physical dimension of 60 mm × 60 mm is utilized as an energy harvester to harvest solar energy from the sunlight. The characteristic of the solar panel is first determined experimentally, followed by the design of a suitable power management circuit to ensure maximum power flow from the solar panel to the energy storage device, hence the electrical load (i.e., wireless sensor node).

5.2.2.1 *Characterization of a Solar Panel*

The experimental setup for the SEH subsystem is constructed to simulate sunlight in the laboratory with a controlled environment. It is difficult to conduct experiments in an outdoor environment where the natural sunlight tends to fluctuate over time. There is no control over the light intensity of the sunshine. Hence, the light intensity and spectrum of the sunlight are emulated with a light source that closely resembles the frequency spectrum of the sunlight. The light source is an OSRAM 300W Ultra Vitalux lightbulb that has the sunlike radiation properties and is widely used in industrial applications [141] for lighting purposes. To vary the solar irradiance of the emulated light, a variable transformer was used to control the electrical input power to the lightbulb. The P-R and P-V curves (see Figures 5.4 and 5.5, respectively) of the solar panel were obtained experimentally through the variation of the solar irradiance level and with different load resistances connected across the solar panel. Throughout the characterization process of the solar panel, the temperature of the solar panel was kept constant by having a constant stream of air blowing across the solar panel.

The P-R curve seen in Figure 5.4 illustrates how the source impedance of the solar panel varies with the changing solar irradiance. As can be seen, the source impedance of the solar panel decreases from 350 to 50 Ω as the light irradiance level increases from 80 to 800 W/m^2. This is due to the higher output current generated by the solar panel at higher solar irradiance. Comparing between the P-R curves of the wind turbine generator and the solar panel given in Figure 2.4 and Figure 5.4, respectively, it is clearly seen that

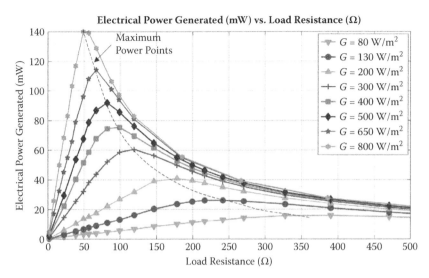

FIGURE 5.4
Power curves of a solar panel over a range of load resistances.

their characteristics are totally different. Unlike the wind turbine generator's characteristic observed in Figure 2.4 where its internal impedance is almost the same for all incoming wind speeds, the internal impedance of the solar panel varies with the changing solar irradiance.

Conversely, it is observed from Figure 5.5 that the MPPT voltage V_{mppt} of the solar panel at which the harvested power is maximum is within a small voltage range of 2.55 to 2.75 V for different solar irradiances. The maximum

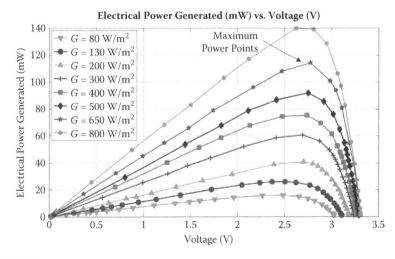

FIGURE 5.5
Power harvested by a solar panel plot against generated voltage for a range of solar irradiance.

electrical power that can be harvested from the solar panel across various solar irradiance levels of 80 to 800 W/m^2 is 16 to 141 mW. By keeping the operating voltage of the solar panel near to V_{mppt}, maximum electrical power is attainable from the solar panel. Other than the complex and energy-hungry MPPT techniques [87–89] like the P&O method and IncCond method, there are several simple and low-power indirect MPPT techniques discussed in Section 5.1 (i.e., FOCV, use of additional pilot solar cell, and CV) that are suitable for small-scale SEH. However, there are some problems associated with the FOCV and pilot solar cell methods as discussed by Dondi et al. [36]. To overcome the problems, the CV method is used in the proposed SEH subsystem to achieve MPPT. With reference to Figure 5.5, the reference MPPT voltage $V_{mppt,ref}$ of the solar panel is set as 2.58 V. At $V_{mppt,ref}$, all the harvested powers of the solar panel are close to its MPPs for a variety of solar irradiance from 80 to 800 W/m^2.

5.2.2.2 Boost Converter with Constant-Voltage-Based Maximum Power Point Tracking (MPPT)

To optimize the SEH subsystem, a DC-DC boost converter is designed to perform MPPT based on the chosen CV scheme. The main functions of the boost converter in the power management circuit of the SEH subsystem (see Figure 5.6) are (1) to step up the low DC voltage output of the solar panel

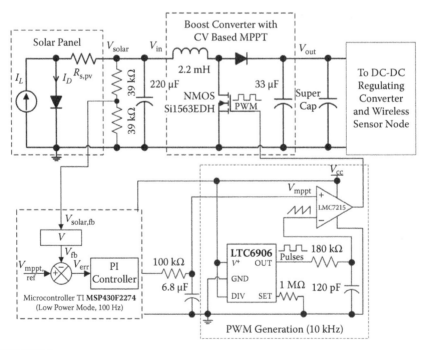

FIGURE 5.6
Schematic diagram of an SEH subsystem.

V_{solar} to charge the energy storage device and (2) to perform near-MPPT so that maximum power transfer takes place. Referring to Figure 5.6, there is a voltage-sensing circuit, essentially a simple resistive voltage divider, to sense and divide the output voltage of the solar panel into two for the microcontroller to process. The power loss in this voltage divider circuit is very small, a few tens of microwatts, and it is quite insignificant as compared to the harvested power on the order of milliwatt level. The feedback voltage signal V_{fb}, obtained from the terminal of the solar panel is compared with the reference voltage signal $V_{mppt,ref}$ in a microcontroller to perform the closed-loop MPPT control of the boost converter via the pulse width modulation (PWM) generation circuit.

The PWM generation circuit, as seen in Figure 5.6, is used to convert the low-frequency PWM control signal, about 100 Hz, generated from the low-power microcontroller to a much higher switching frequency, 10 kHz, so that smaller filter components can be used in the boost converter to miniaturize the overall SEH subsystem. Depending on the voltage, hence the energy storage level of the supercapacitor, V_{cap} or V_{out}, the output voltage of the solar panel V_{solar} is manipulated to transfer maximum power to the supercapacitor by adjusting the duty cycle of the PWM gate signal of the boost converter such that V_{solar} is as close as possible to $V_{mppt,ref}$, the reference MPPT voltage of 2.58 V at which the harvested power is near its maximum. As the energy level of the supercapacitor increases or decreases, the output voltage of the boost converter V_{out} varies with the supercapacitor's voltage. However, at the input voltage of the boost converter V_{in}, it is fixed to the solar panel's reference MPPT voltage $V_{mppt,ref}$ of 2.58 V; hence, the optimal duty cycle of the boost converter D_{opt}, which ensures near MPPT takes place for the solar panel, is expressed as

$$D = 1 - \frac{V_{out}}{V_{in}} \tag{5.4}$$

$$D_{opt} = 1 - \frac{V_{cap}}{V_{mppt,ref}} \tag{5.5}$$

To experimentally verify the concept of a constant voltage approach to perform MPPT for small-scale SEH, a set of experiments was conducted. During the experiments, the MPPT capability of the designed boost converter circuitry was tested for different solar irradiance, and the experimental results are shown in Figure 5.7.

Initially, for the first 20 s, it is observed in Figure 5.7 that the boost converter was not controlled to operate the SEH subsystem at its MPP. After this, the MPP tracker utilizes the closed-loop proportional integral (PI) controller to manipulate the duty cycle of the boost converter according to Equation 5.4, which in turn controls the input voltage of the boost converter towards the optimal reference voltage value of 2.58 V. Once the MPP of the power curve for a wind speed of 80 W/m² was reached, the closed-loop MPP tracker control the boost converter to maintain power harvested from the solar panel

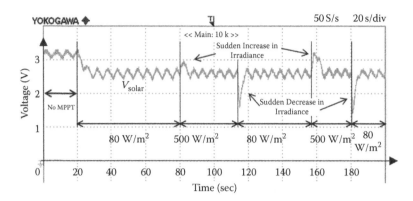

FIGURE 5.7
Performance of a boost converter with CV-based MPPT under varying solar irradiance.

for all the other MPPs occurring at different solar irradiance. As can be seen in Figure 5.7, there were several sudden variations of irradiance levels at different intervals. The closed-loop PI feedback control is able to correct the output voltage of the solar panel back to the MPPT voltage of 2.58 V. This shows that the changing operating conditions of the deployed environment is well taken care of by the proposed closed-loop CV scheme of the designed boost converter.

The designed boost converter with a CV-based MPPT approach has been demonstrated to yield capability in extracting maximum power from the solar panel, but this comes at the expense of additional power losses in the converter. The efficiency of the boost converter η_{conv} is determined based on the function of its output load power P_{load} over its input DC power P_{dc}. Taking the target deployment area with average solar irradiance of 80 W/m^2 as an example, the efficiency of the converter is calculated to be as follows:

$$\eta_{conv} = \frac{P_{out}}{P_{in}} * 100\% = \frac{V_{out}^2 / R_{load}}{V_{in} I_{in}} * 100\% \tag{5.6}$$

$$= \frac{3.98 V^2 / 1200\Omega}{2.58 V * 5.6 \, mA} * 100\% = 91\%$$

For all other illuminations and resistance loadings, the efficiencies of the boost converter are calculated using Equation 5.6 to be between 80% and 92%, and the computed results are shown in Figure 5.8. It is observed from Figure 5.8 that the efficiency of the boost converter generally drops as the irradiance level increases. This is due to increasing current output from the solar panel when light irradiance increases. The increased current contributes to ohmic losses, hence causing a lower efficiency level to be attained. Referring to Figure 5.8, it can be seen that even for a low solar irradiance condition where the power harvested is small (around 14.4 mW), the boost converter is still able to achieve a reasonably good efficiency of 91%. This exhibits the ability

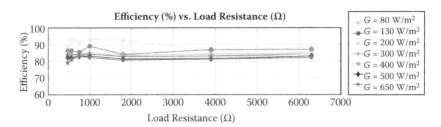

FIGURE 5.8
Efficiency of an MPPT boost converter for various solar irradiance.

of the DC-DC boost converter to attain high efficiency in a condition of very low power rating.

5.2.2.3 Performance of SEH Subsystem

To evaluate the performance of the SEH subsystem with MPPT and without MPPT, the electrical load was changed from a pure resistor to an energy storage device (i.e., a supercapacitor). Unlike the resistor with constant resistance, the impedance of the supercapacitor is always changing in accordance with its energy storage level. During its charging process, the dynamic response of the supercapacitor is of importance to be considered in the design of the boost converter to ensure constant MPPT operation is achieved. Figures 5.9a and b show two separate experiments being conducted on the SEH subsystem with MPPT and without MPPT, respectively, at a solar irradiance of 80 W/m^2.

Referring to Figure 5.9a, it is observed that the SEH subsystem with MPPT charged the supercapacitor from a voltage level of 2.815 to 3.13 V in 500 s, while in Figure 5.9b, the supercapacitor was charged from 2.84 to 2.975 V in 500 s without any MPPT control. The total amount of energy stored in the supercapacitor was determined to be around 4.68 J, while its counterpart was only 1.96 J. It is obvious that SEH with MPPT control outperformed its counterpart without MPPT control. The amount of energy harvested by the SEH subsystem with MPPT was 2.39 times more than the amount of energy

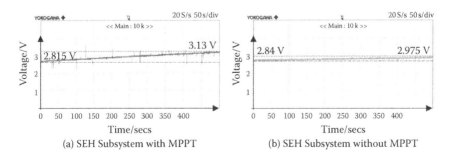

FIGURE 5.9
Performance of an SEH subsystem with MPPT and without MPPT for charging a supercapacitor.

harvested without MPPT. This proves that MPPT is beneficial in maximizing the amount of energy that can be harvested from the solar panel.

5.2.3 Hybrid Solar and Wind Energy Harvesting System

In this HEH research, the outputs of the WEH and SEH subsystems were connected in parallel so that more power could be delivered to the electrical load. The expected outcome of this hybrid topology is the summation of the harvested powers of the WEH and SEH subsystems as shown in Figure 5.3. The two diodes as shown in Figure 5.3 connect the outputs of the WEH and SEH subsystems together, and they play a critical role in their successful combination. The role of the diode is to prevent any current flow from the output of either boost converter to the other subsystem as shown in Figure 5.10. As such, simultaneous charging from both energy sources is accomplished. In addition, with these blocking diodes, the proposed HEH topology is able to carry out MPPT operations in the WEH and SEH subsystems. The WEH subsystem uses the resistor emulation (RE) technique, while the SEH subsystem uses the CV technique.

Referring to Figure 5.10a, when the WEH subsystem is operating in MPPT mode, it is observed that the SEH subsystem appears as an infinite resistance to the WEH subsystem. The effective load experienced by the WEH subsystem is just simply the energy storage device. Similarly, for the SEH subsystem seen in Figure 5.10b, it sees the WEH subsystem as an infinite resistance,

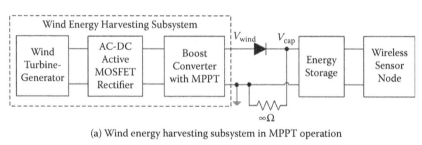

(a) Wind energy harvesting subsystem in MPPT operation

(b) Solar energy harvesting subsystem in MPPT operation

FIGURE 5.10
Efficiency plot of a diode block under varying solar irradiance and resistance loadings.

which is equivalent to an open-circuit condition. This is due to the diode at the output of the SEH subsystem that prevents current flow from the WEH subsystem as well as the energy storage. The CV MPPT technique used in the SEH subsystem fixes the input voltage of its boost converter while varying the output voltage of the boost converter according to the voltage, hence the energy storage level of the supercapacitor. The MPPT operations of the WEH and SEH subsystems are independent.

In this HEH system, the two diodes play an important role in insolating the two subsystems so that maximum power is transferred from the energy source to the load. However, some amount of power loss would be incurred in the HEH system. The efficiency of the diode block is on average about 93% over a range of solar irradiance and resistance loading conditions. This positive outcome is attributed to the very low power loss in the Schottky diode, where the voltage drop across the Schottky diode is low, in the range of 0.15 to 0.25 V. As such, it is viable to employ the diode block in the designed HEH system to ensure MPPT operations are performed by the WEH and SEH subsystems. Other than the power loss in the boost converter and the diode block, another investigation was carried out to determine the power loss in the associated control, sensing, and PWM generation electronic circuits. The supply voltage of the electronic circuits provided by a voltage regulator is 3 V. Based on the current requirement of each individual component in the sensing and processing circuits, a summary of the total power consumption of the electronic circuits was calculated and tabulated.

As can be seen in Table 5.1, the total power consumed by the associated control, sensing, and PWM generation electronic circuits of the WEH and SEH subsystems was calculated to be 1.135 mW. As compared to the near-maximum power harvested by the optimized HEH system seen in Figure 5.11, which ranged from a few tens to hundreds of milliwatts, the power consumption of the designed electronic circuits was very low and insignificant. Taking the target deployment area with average wind speed of 4 m/s and average solar irradiance level of 80 W/m^2 as an example, it can be seen from Figure 5.11 that the total power harvested by the HEH system was 34 mW. The power loss of 1.135 mW was only a small fraction, around 5%, of the harvested power. Even for very low wind speed of 2.3 m/s and low solar irradiance of

TABLE 5.1

Power Consumption of Associated Control, Sensing, and PWM Generation Electronic Circuits

Component	Subsystem	Qty	Current/μA	Power/μW
Comparator	WEH	3	7 * 3 = 21	63
2 op-amps	WEH	1	90	270
Comparator	SEH	1	7	21
Voltage divider	SEH	—	—	85
Oscillator	Both	1	12	36
Wireless sensor node/μC	Both	1	220	660

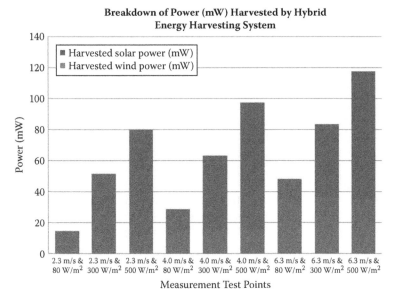

FIGURE 5.11
Power harvested by the hybrid wind and solar energy harvesting system.

$80 \text{ W}/\text{m}^2$, the harvested power of 17 mW is still more than enough to sustain the operation of the wireless sensor node.

5.2.4 Experimental Results

The proposed concept of a self-powered HEH wireless sensor node using an efficient power management circuit, as illustrated in Figure 5.12, has been implemented into a hardware prototype tested in the laboratory environment. A photograph of the developed HEH wireless sensor node is depicted in Figure 5.13. Several tests were conducted during the experiments to validate the performance of the optimized HEH system using a CV-based MPPT scheme in sustaining the operation of the wireless sensor node.

5.2.4.1 Performance of the Hybrid Energy Harvesting (HEH) System

The experimental tests were conducted in accordance with the winter condition of the deployment ground illustrated in Section 5.2.1 where the average wind speed and average solar irradiance level were given as 4 m/s and $80 \text{ W}/\text{m}^2$, respectively. The HEH system is designed to sustain the wireless sensor node to transmit the sensed data (i.e., temperature and wind speed) every second. A 1.5-F, 5.5-V supercapacitor is used to store the harvested energy from the HEH system. Several experimental tests were conducted on the hardware prototype of the HEH wireless sensor. The experimental results shown in Figure 5.14 are used to differentiate the performance of the WEH

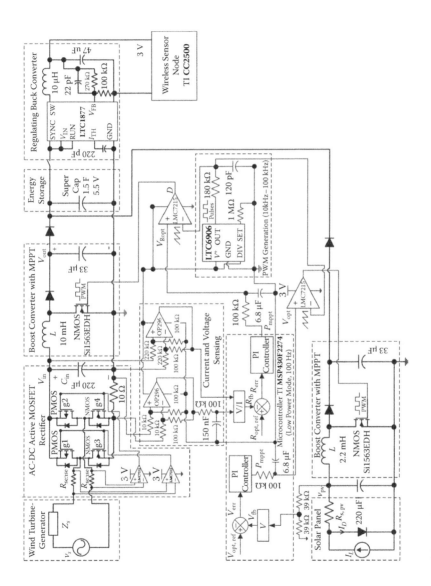

FIGURE 5.12
Schematic diagram of the wind-powered wireless sensor node augmented with an SEH subsystem.

FIGURE 5.13
Hardware prototype of a hybrid energy harvesting wireless sensor node.

and SEH subsystems and their MPPT schemes in powering the electrical load consisting of a supercapacitor, sensing and control circuits and wireless sensor node.

In region A of Figure 5.14, the electrical load is at first powered with the SEH subsystem with a constant voltage-based MPPT scheme. It can be seen from Figure 5.14 that the output voltage of the solar panel V_{solar} was fixed at its MPPT reference voltage $V_{mppt,ref}$ of 2.58 V. This tracking profile verifies that the SEH subsystem was operating near its MPP. During this time duration of 90 s, the 1.5-F supercapacitor's voltage V_{cap} was charged from 3.1 to 3.35 V, and around 13 mW of electrical power was harvested by the SEH subsystem at an average solar irradiance level of 80 W/m². However, in region B1 of Figure 5.14, when the SEH subsystem was disconnected from the load, the output voltage of the supercapacitor V_{cap} fell back to 3.1 V in 100 s. In region C, V_{cap} continues to decrease at a slower rate. This is because solar energy was again harvested but not at the MPP. The output voltage of the solar panel V_{solar} is seen in Figure 5.14 as 2.94 V, which is higher than the MPPT voltage reference of 2.58 V. Without MPPT control for the SEH subsystem, the harvested solar energy was not able to sustain the entire unit.

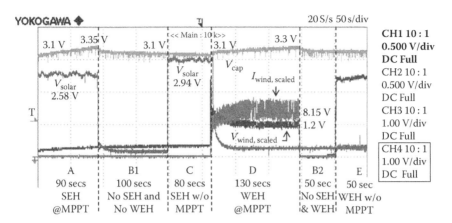

FIGURE 5.14
Operation of a sensor node under various powering schemes.

In region D of Figure 5.14, the WEH subsystem with an RE-based MPPT scheme was employed. It is observed from Figure 5.14 that the output voltage of the supercapacitor V_{cap} increased from 3.1 to 3.35 V in 130 s. Referring to Figure 5.14, the DC voltage of the wind turbine generator $V_{wind,scaled}$ (scaled DC voltage of wind turbine generator: multiply experimental value by 2.2 to get actual voltage), is 1.2 V, while its current $I_{wind,scaled}$ (scaled voltage equivalent of DC current of wind turbine generator: multiply experimental value by 0.01 to get actual current in amperes) is 8.15 mA. Hence, the calculated resistance of $R_{em} = \frac{1.2V}{8.15 \text{ mA}} = 147\ \Omega$ was equivalent to the emulated resistance of $150\ \Omega$. This equivalence shows that the WEH subsystem was operating at near its MPP, and it was able to harvest around 9.5 mW of electrical power at an average wind speed of 4 m/s. However, in region B2 of Figure 5.14, when the WEH subsystem was disconnected from the load, the output voltage of the supercapacitor V_{cap} falls back to 3.1 V in 50 s. In region E, it is observed that the supercapacitor's voltage continued to decrease even though wind energy was being harvested again but not at MPPT. The reason is that the voltage and current at the input of the boost converter of the WEH subsystem without MPPT was no longer controlled, resulting in much less harvested power. As such, MPPT was important for optimal and effective EH performance of the WEH and SEH subsystems.

Another set of experiments was conducted to evaluate the performance of the HEH system in comparison with the WEH and SEH subsystems individually. The outcome of the performance comparison between a single-source-based EH system and a multiple-source-based energy harvesting system is illustrated in Figure 5.15.

Referring to Figure 5.15, it can be observed from regions A and C that the single-source EH systems (i.e., SEH and WEH) operating in MPPT mode were able to charge the 1.5-F supercapacitor as well as power their associated sensing and control circuit and wireless sensor node. The electrical power

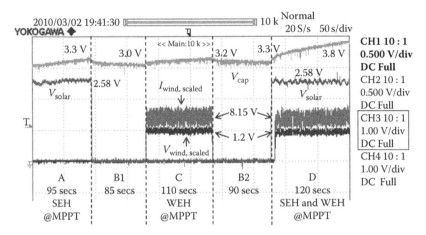

FIGURE 5.15
Comparison between single ambient energy source harvesting and hybrid energy harvesting.

harvested by the SEH and WEH subsystems were computed to be 13 and 9.5 mW, respectively. In region D of Figure 5.15, it can be seen that the HEH system was harvesting simultaneously from both the solar and wind energy sources, so the supercapacitor's voltage increased at a much faster rate, from 3.3 to 3.8 V in about 120 s. Based on the net energy accumulated in the supercapacitor, the electrical power harvested by the HEH system, operating in MPPT mode, to charge the supercapacitor was calculated to be around 22.2 mW. This calculated output power of the HEH system is almost equivalent to the summation of the electrical power output from the SEH and WEH subsystems, that is, 12.65 and 9.85 mW, respectively, of 22.5 mW. It is clearly demonstrated that HEH has tangible benefits over single renewable ambient source EH. The supercapacitor charges much faster compared to single ambient source EH because of simultaneous charging from both solar and wind energy. As a result, more energy is harvested in a shorter span of time, allowing duty cycling operations of the wireless sensor node to increase.

5.2.4.2 Power Conversion Efficiency of the HEH System

The HEH system consists of two main EH subsystems, namely the WEH and SEH subsystems. In each subsystem, there are stages of efficiency loss, and these should be examined in greater detail to determine the amount of power lost per stage and thus determine the performance of each stage. A line diagram to illustrate the input and output power available for each stage of both subsystems is provided in Figure 5.16. It is intended to give a clearer picture of understanding how power is distributed at each power conversion stage.

Referring to Figure 5.16, the line diagram of the WEH subsystem starts from the input with a wind speed of 4 m/s where 18 mW of raw electrical

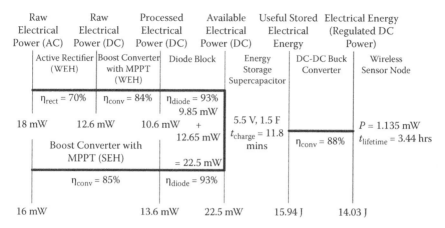

FIGURE 5.16
Line diagram of power distributed in the hybrid energy harvesting system.

power (AC) is generated at the output of the wind turbine generator. Using the designed power management circuit, which included an active rectifier and a boost converter with MPPT, the harvested AC power of 18 mW was converted into 10.6 mW electrical DC power. Similarly, the SEH subsystem harvested around 13.6 mW of electrical DC power from an input of 16 mW of raw electrical power after the power loss in the boost converter. Compared with the WEH subsystem, the SEH subsystem suffered comparatively less efficiency loss as it had only one conversion stage. The effective power harvested from both subsystems was input into the diode block with a high efficiency of 93% and the combined output power was calculated to be 22.5 mW.

Since the supply voltage of the associated control, sensing, and PWM generation electronic circuits was 3 V, the maximum amount of energy stored in the 1.5-F supercapacitor, starting from 3 to 5.5 V was 15.94 J. To fully charge the supercapacitor, the required charging time t_{charge} is computed to be 11.8 min when 22.5 mW of electrical power was supplied. A DC-DC buck converter with efficiency of 88% was used to regulate the supply voltage at 3 V. This amounted to 14.03 J of useful electrical energy stored in the supercapacitor. Referring to Figure 5.16, it is illustrated that the HEH system with its fully charged supercapacitor was able to sustain the operation of the wireless sensor node for 3.44 hr. The line diagram seen in Figure 5.16 clearly illustrates that the amount of power harvested, 22.5 mW, by the HEH system was nearly twice that of the individual WEH or SEH subsystem. Hence, the operational lifetime of the sensor node powered by the HEH system with the MPPT scheme was twice of the sensor node powered solely by either the SEH or WEH subsystem, hence making the HEH system with the MPPT scheme a viable solution for extending the lifetime of the wireless sensor network (WSN).

5.2.5 Summary

In this section, a hybrid of WEH and SEH (HEH) scheme was proposed for outdoor applications. The existing WEH subsystem was augmented with the developed SEH subsystem to extend the operational lifetime of the wireless sensor node. The proposed HEH topology used diodes to enable MPPT operations to be carried out in the WEH and SEH subsystems. The WEH subsystem uses the RE technique, while the SEH subsystem uses the constant voltage technique. This ensures simultaneous optimal charging of the supercapacitor as well as powering of the wireless sensor node. Experimental tests were conducted in accordance with the winter condition of the deployed ground where the average wind speed and average solar irradiance level were given as 4 m/s and 80 W/m^2, respectively. The electrical power harvested by the SEH and WEH subsystems of the HEH system was 13 and 9.5 mW, respectively. The total electrical power harvested by the optimized HEH system was 22.5 mW, which was almost two times higher than the highest of the single-source-based EH method.

5.3 Composite Solar, Thermal (S+T) Energy Sources

The concept of HEH from two readily available energy sources to augment the life span of a wireless sensor node powered by single-source EH was validated in Section 5.2. Many EH systems have already been developed, and those reported in the literature are mostly for outdoor applications where solar energy is plentiful. Very few research studies [143–144] have discussed the indoor EH from an ambient light source, which has serious challenges to resolve. In this section, the concept of EH from two readily available indoor energy sources—ambient light energy from artificial lighting and thermal energy from machine heat—to sustain low-power electronic remote sensors for supervisory and alarm system is proposed.

There are two main challenges associated with indoor EH: (1) ambient energy sources in the indoor environment are very weak, so the harvested power is much lower than that of the outdoor condition; and (2) availability of energy sources is dependent on the indoor environmental conditions. According to Randall et al. [30], it has been shown that the light intensity of artificial lighting conditions found in hospitals and office buildings is only a small fraction of the light intensity for outdoor sun of 100 to 1000 W/m^2. As such, the power density of an amorphous solar cell with efficiency of less than 10% under indoor light intensity of less than 10 W/m^2 is significantly reduced to 100 μW/cm^2 [145]. This is even more challenging when indoor EH is available for a limited period of time. The ambient energy sources can be intermittent and inconsistent at times depending on the indoor environmental conditions. Take, for example, if indoor lighting is only available during office hours and there is almost complete darkness for the rest of the day. Because of these challenges related to the weak and uncertain energy sources, EH in an indoor environment from a single energy source might not be adequate to sustain the operation or even to enhance the performance of the miniaturized wireless sensor nodes over the lifetime.

Hybrid energy harvesting has been recently discussed in the literature [35, 126–129] as a potential micropower supply solution to minimize the size of the energy supply as well as to extend the operational lifetime of the wireless sensor node. Researchers have considered a number of methods to combine different small-scale EH sources, and these methods can be classified into four main categories as described in Section 5.2. For the first three HEH methods discussed in the literature, each EH source is required to have its own unique power management unit (i.e., DC-DC converter) and associated circuitry to condition the power flow from the energy source to its output load. As more energy sources are combined, the number of power management units for each individual energy source increases, hence more components are needed and larger volumetric size, higher power losses, and costs are incurred. This is not the case for the fourth HEH method proposed in this chapter. The proposed HEH system requires only one power management unit to condition the combined output power harvested from the solar and thermal energy sources.

Emphasis of this section is placed on enhancing the performance of the wireless sensor node deployed in the challenging indoor context using HEH from solar and thermal energy sources. A near MPPT technique is explored for the HEH system to maximize power transfer from the hybrid energy sources to the sensor node. The rest of the section is organized as follows: Section 5.3.1 provides an overview of the indoor energy sources. Section 5.3.2 illustrates the details of the indoor SEH subsystem, while Section 5.3.3 explains the TEH subsystem. Section 5.3.4 discusses how the SEH and TEH subsystems are combined directly in parallel configuration. Following that, the optimized HEH system using a single power management unit for an indoor wireless sensor node prototype is illustrated in Section 5.3.5; the section ends with a summary and discussion in Section 5.3.6.

5.3.1 Overview of Indoor Energy Sources

The characteristics and performances of the renewable energy sources available in outdoor environmental conditions are very different from those found in indoor industrial, commercial, and biomedical environments. Within an enclosed environment like offices, hospitals, factories, and so on, the energy sources are generally generated by some artificial means. Table 5.2 shows a summary of the indoor energy sources and their characteristics.

As seen in Table 5.2, the artificial energy sources found in an indoor environment are generated from electrical appliances such as artificial lighting, air conditioners, machines, and so on, and human movement. Each of these appliances and human bodies has its own primary purpose. Artificial lighting and an air conditioner, for example, are primarily used to generate light

TABLE 5.2

Characteristics of Indoor Energy Sources

Energy Source	Characteristics	Comments
Solar	Intermittent	Illumination from artificial lighting during office hours
Wind	Continuous	Air circulation from air conditioner and electric fans
Thermal	Continuous/intermittent	Thermal gradient between body, machine heat, and ambient
Vibration	Intermittent	Vibration from machine and human motion during walking, running, etc.

for visual function and air circulation for cooling purposes. During the normal operation of these electrical appliances and human movements, some stray energies are generated for EH. The availability of the artificial energy sources seen in Table 5.2 is largely dependent on the operational schedules of the electrical appliances and the lifestyles of the human beings. Some energy sources, such as solar and vibration, are only available according to the intermittent operating profile of the artificial lighting, machine operation, and human movement. Conversely, the wind and thermal energy sources exist in continuous form as the air conditioners and machineries of indoor industrial, commercial, or biomedical environment are operated at a duty cycle close to 100% of the time.

Apart from the issue of availability of energy sources, another concern is power harvesting throughput. The power densities of various EH technologies per unit square centimetre or cubic centimetre are recorded in Table 5.3. Under indoor conditions, it can be observed that all the artificial energy sources become very weak. The power density of a solar panel, for example, with efficiency of less than 10% under indoor solar irradiance of 10 W/m^2 is only 100 μW/cm^2 as compared to 10 mW/cm^2 under the outdoor standard testing condition (STC). Table 5.3 shows that the average power that can be harvested by all the artificial energy sources is 10 to 100 times lower than that of the outdoor ambient energy sources. As such, these weak and

TABLE 5.3

Performance of Energy Harvesters under Indoor Conditions

Energy Harvester	Power Densities	
	Indoor Condition	Outdoor Condition
Solar panel	100 μW/cm^2 at 10 W/cm^2	10 mW/cm^2 at STC
Wind turbine generator	35 μW/cm^2 at <1 m/s	3.5 mW/cm^2 at 8.4 m/s
Thermoelectric generator	100 μW/cm^2 at 5°C gradient	3.5 mW/cm^2 at 30°C gradient
Electromagnetic generator	4 μW/cm^3 at human motion-Hz	
	800 μW/cm^3 at machine-kHz	

uncertain indoor energy sources pose significant challenges for EH from a single energy source for sustaining the operation of the wireless sensor nodes over the entire lifetime.

To enhance the performance of the wireless sensor node in an indoor environment, HEH from solar and thermal energy sources is proposed in this section. Among the various artificial energy sources, solar and thermal energy sources share similar high power densities, as can be seen in Table 5.3. In addition, the undesirable intermittent solar energy sources found in offices or factory lighting can be supplemented by the continuous thermal energy supply from waste heat generated by machinery. Whenever solar energy is not available, instead of solely relying on the energy stored in the onboard energy storage devices proposed by Nasiri et al. [143] and Hande et al. [144], an alternate approach is to harvest from the readily available thermal energy source to continue powering the operation of the wireless sensor node before exhausting the energy stored in the energy storage device. Another key characteristic of the proposed HEH system is that the HEH system is able to harvest simultaneously from both energy sources whenever they are available instead of harvesting from an individual energy source at one time [126–129], hence the performance of the indoor wireless sensor node can be enhanced.

5.3.2 Indoor SEH Subsystem

The solar panel used in this section is specially selected for use under indoor conditions (i.e., artificial lighting from fluorescent lamps) and at room temperature (relatively less variation of temperature in an indoor environment than outdoors [30]). The physical size of the chosen solar panel is 55 × 30 × 1 mm (2.1 × 1.18 × .04) and its cross-sectional area A can be calculated to be around 16.5 cm². At very low light illumination, for example, $G = 380$ lux ($\approx 380/120 = 3.17$ W/m² [30]), the open-circuit voltage V_{oc} and short-circuit current I_{sc} are measured to be 4.14 V and 60 μA, respectively. The output voltage and current obtained at MPPT points are $V_{mppt} = 3.5$ V and $I_{mppt} = 51.44$ μA, respectively. For the given technical characteristics of the solar panel, the corresponding solar panel's efficiency can also be determined using the following equation:

$$\eta = \frac{P_{pv}}{G * A} * 100\% \tag{5.7}$$

Based on Equation 5.7, the efficiency of the solar panel can be calculated to be around 3.4%, which is relatively lower than the outdoor solar panel [145]. Due to the low efficiency of the solar panel in the indoor condition, the power harvested is also low; hence, it is necessary to optimize the indoor SEH subsystem in order to maximize the power harvested from the solar panel. Further investigations were carried out on the solar panel to investigate its performance in different lighting conditions. Figures 5.17 and 5.18 show the indoor solar panel's P-V and P-R curves at different lux illuminations. Both

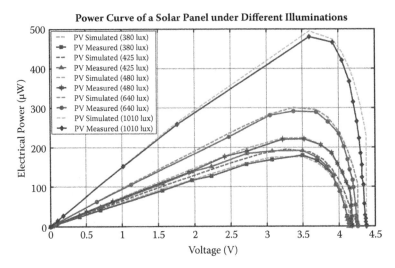

FIGURE 5.17
P-V curves of a solar panel at different lux conditions.

the power curves of the solar panel in relationship with the output voltage (P-V) and the load resistance (P-R) under a range of loads from short circuit to open circuit were generated at five different lighting conditions ranging from 380 to 1010 lux.

Taking into consideration the parameters extracted for the solar panel, the model of the solar panel expressed by Equation 5.3 has been simulated, and the simulation results are recorded in Figure 5.17. The simulated P-V curves

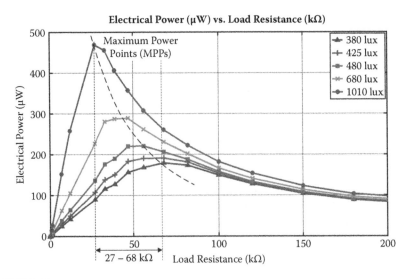

FIGURE 5.18
P-R curves of a solar panel at different lux conditions.

are verified with the measured P-V curves as seen in Figure 5.17 for varying solar irradiance conditions, performed by a characterization setup based on a fluorescent light source. Figure 5.17 shows that all of the power curves peak near the solar panel output voltage of 3.6 V. Conversely, for the power curve (P-R) plotted in Figure 5.18, it can be observed that the MPPs of the solar panel vary between 27 and 68 kΩ. As such, by setting the output voltage of the solar panel fixed at 3.6 V, maximum output power can be harvested from the solar panel under different solar irradiance. Within the indoor lighting conditions of 380 to 1010 lux, the maximum electrical power that the solar panel can harvest ranges from 180 to 480 µW, respectively.

5.3.3 Thermal Energy Harvesting Subsystem

In the TEH subsystem, a miniaturized thermoelectric generator (TEG) housed in the thermal energy harvester is used for converting thermal energy into electrical energy. The thermal energy, generated from the heat source at a certain high temperature of T_H, is channelled through the enclosed TEG via a thin film of thermally and electrically conductive silver grease between them to the heat sink. The residual heat accumulated in the heat sink is then released to the surrounding ambient air at a lower temperature T_C. An equivalent thermal circuit model of the thermal energy harvester that illustrates its thermal and electrical characteristics is provided in Figure 5.19.

As observed in Figure 5.19, the temperature difference ΔT_{TEG} across the junctions of the TEG is lower than the temperature gradient, $\Delta T = T_H - T_C$, that is externally imposed across the thermal energy harvester. This is because

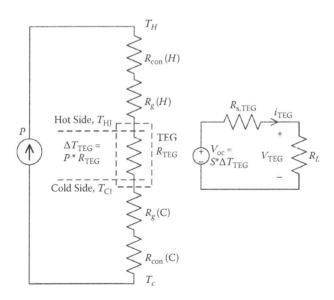

FIGURE 5.19
Equivalent electrical circuit of the thermal energy harvester.

of the thermal contacts and thermal grease resistances residing in the cold and hot sides of the thermal energy harvester, that is, $R_{con(H)}$, $R_{con(C)}$ and $R_{g(H)}$, $R_{g(C)}$, respectively. To minimize this negative effect, the thermal resistance R_{TEG} of the TEG is made to be as high as possible; conversely, the rest of the thermal resistances of the thermal energy harvester are designed to be as small as possible. Taking these design considerations into account, the miniaturized thermal energy harvester, having a physical size of $20 \times 20 \times 20$ mm, is designed in such a way that most of the heat is channelled through the TEG in order to maximize TEH.

Analysis and characterization works were conducted on the designed thermal energy harvester to evaluate the performance of the TEH subsystem in powering the wireless sensor node. According to Seebeck's effect, the open-circuit voltage V_{oc} of the TEG enclosed in the thermal energy harvester, which is composed of n thermocouples connected electrically in series and thermally in parallel, is given as

$$V_{oc} = S * \Delta T = n * \alpha (T_H - T_C) \tag{5.8}$$

where α and S represent Seebeck's coefficient of a thermocouple and a TEG, respectively. When connecting a load resistance R_L electrically to the TEG via the thermal energy harvester as shown in Figure 5.19, an electrical current I_{TEG} flows in accordance to the applied temperature difference ΔT, which is given as

$$I_{TEG} = \frac{V_{oc} - V_{TEG}}{R_{s,TEG}} = \frac{n * \alpha (T_H - T_C) - V_{TEG}}{R_{s,TEG}} \tag{5.9}$$

where $R_{s,TEG}$ is the internal electrical resistance of the TEG. Based on the current-voltage (I-V) characteristic of the TEG described in Equation 5.9, the output power $P_{TEG}(V_{TEG})$ delivered by the TEG to the load R_L can be determined. By substituting I_{TEG} with Equation 5.9, the electrical power $P_{TEG}(V_{TEG})$ harvested by the thermal energy harvester is derived as a function of its output voltage V_{TEG}, which is expressed as

$$\begin{aligned} P_{TEG}(V_{TEG}) &= V_{TEG} * I_{TEG} \\ &= \frac{V_{TEG} * n * \alpha^* (T_H - T_C) - V_{TEG}^2}{R_{s,TEG}} \end{aligned} \tag{5.10}$$

Based on the technical specifications provided for the Thermo Life TEG [91], the TEG used in this section is made up of 5200 thermocouples, and each thermocouple has a Seebeck's coefficient α of 0.21 mV/K. For a given temperature difference $\Delta T = T_H - T_C$, between 5 and 10 K, the model illustrated by Equation 5.10 is simulated for different TEG's output voltage V_{TEG}, and the simulation results are presented in Figure 5.20. Experiments were carried out to characterize the TEG by applying a temperature difference between

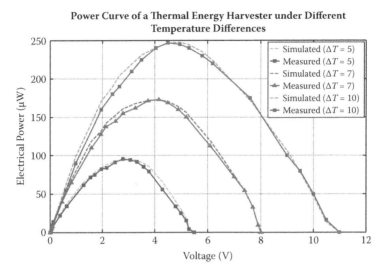

FIGURE 5.20
P-V curves of a thermoelectric generator at different thermal gradients.

the thermal contact faces, and both the electrical output voltage and current with different loads connected were measured. This experiment was repeated for various temperature differences in the range between 5 and 10 K, and the experimental results are shown in Figure 5.20.

Referring to the power curve (P-V) shown in Figure 5.20, it can be seen that the simulation results obtained using the model expressed in Equation 5.10 are comparable to the measurement results collected from the characterization of the thermal energy harvester under varying temperature differences. Figure 5.20 shows that the maximum obtainable power for each thermal gradient corresponds to an output voltage of the thermal energy harvester, that is, $P_{mppt, \Delta T=5K} = 96 \ \mu$W at 2.8 V, $P_{mppt, \Delta T=10K} = 247 \ \mu$W at 4.5 V, and so on. This is unlike the indoor SEH case, whereby all their power curves (P-V) peak near a particular output voltage of the solar panel (see Figure 5.17).

However, this is not the case for another power curve (P-R) plotted for SEH and TEH as shown in Figures 5.18 and 5.21, respectively. It can be observed that the MPPs of the solar panel vary between the load resistances of 27 to 68 kΩ, whereas the MPPs of the thermal energy harvester are fixed at the internal impedance of the thermal energy harvester of 82 kΩ. Figure 5.21 shows that when the load resistance matches the source resistance of the thermal energy harvester, the harvested power is always maximum for different temperature differences. Because of that, it can be concluded from both power curves in Figures 5.18 and 5.21 that no common MPPT approach exists between the SEH and TEH subsystems.

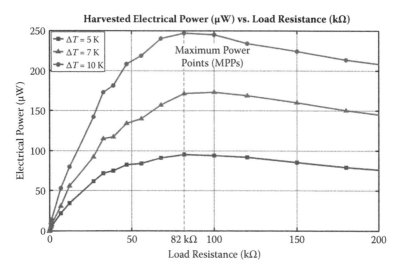

FIGURE 5.21
P-R curves of a thermoelectric generator at different thermal gradients.

5.3.4 HEH from Solar and Thermal Energy Sources

5.3.4.1 *Characteristics of a Solar Panel and Thermal Energy Harvester Connected in Parallel*

For the HEH approach proposed in this section, the terminal output voltages of the solar panel and the thermal energy harvester V_{pv} and V_{TEG}, respectively, are directly connected to the load, each via a Schottky diode to block reverse-biased current flow. An overview of the equivalent electrical circuit of the hybrid energy harvester is shown in Figure 5.22.

According to Figures 5.17 and 5.20, the output voltages of the two energy sources are not that low, typically a few volts, hence the series energy sources configuration is not used to step up the voltage across the load V_{R_L}. Instead, a parallel energy source configuration, $V_{R_L} = V_{pv} = V_{TEG}$, is employed to produce more current flows, that is, $I_{R_L} = I_{pv} + I_{TEG}$. The power harvested from the solar panel $P_{pv}(V_{pv})$ and the thermal energy harvester $P_{TEG}(V_{TEG})$, expressed by Equations 5.3 and 5.10, respectively, are summed together to

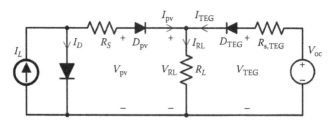

FIGURE 5.22
Equivalent electrical circuit of the proposed hybrid energy harvester.

power the load. The electrical power throughput of the hybrid energy harvester $P_{HEH}(V_{R_L})$ as a function of its output voltage V_{R_L} is thus given by

$$P_{HEH}(V_{R_L}) = |P_{pv}(V_{R_L})| + |P_{TEG}(V_{R_L})|$$

$$\approx \left| V_{R_L} * I_{sc,pv} \right| - \left| V_{R_L} * I_o \left[exp \left(\frac{V_{R_L}}{n_s k T_c / q} \right) \right] \right| + \left| \frac{V_{R_L} * V_{oc,TEG} - V_{R_L}^2}{R_{s,TEG}} \right|$$

$$(5.11)$$

Based on the technical specifications of the solar panel and the thermal energy harvester given in Sections 5.3.2 and 5.3.3, respectively, the harvested power expression of the hybrid energy harvester, as expressed by Equation 5.11, is simulated over a range of output voltages V_{R_L} for solar irradiance and temperature differences that correspond to the solar panel's short-circuit current $I_{sc,pv}$ and the thermal energy harvester's open-circuit voltage $V_{oc,TEG}$. A set of the simulation and experimental results extracted under the minimum (380 lux and $\Delta T = 5$ K) and the maximum (1010 lux and $\Delta T = 10$ K) power harvesting conditions is shown in Figure 5.23.

Referring to Figure 5.23, it can be observed that the measured power curve (Measured S+T) of the hybrid energy harvester is the result of summing the individual power curves, that is, solar panel [Solar (S)] and thermal energy harvester [Thermal (T)], being superimposed into Figure 5.23 minus the negligible small power loss in the Schottky diodes. At MPPT voltage $V_{R_L,mppt}$ of 3.6 V, the output voltages of the solar panel and thermal energy harvester are slightly higher than $V_{R_L,mppt}$ such that the two isolation diodes are conducting in the forward bias condition, hence with reference to Figure 5.23, it can be seen that the hybrid energy harvester can generate power at the minimum of 252 μW ($P_{pv} = 167$ μW, $P_{TEG} = 85$ μW) and at the maximum of 693 μW ($P_{pv} = 466$ μW, $P_{TEG} = 227$ μW). In addition, Figure 5.23 shows that when $V_{R_L} \geq V_{pv}$, the solar panel operates in the open-circuit mode; therefore, no solar power is harvested. This situation happens to the thermal energy harvester as well if $V_{R_L} \geq V_{TEG}$ (3.6 V).

Another observation seen in Figure 5.23 is that the simulated waveforms (Simulated S+T) based on the model expressed by Equation 5.11 and the measured waveforms (Measured S+T) obtained from experiments are quite similar. The positive outcome of this observation verifies the expression model derived in Equation 5.11, which can then be used to determine the electrical power throughput of the hybrid energy harvester P_{HEH} to sustain the operational lifetime of the wireless sensor node. More analysis and characterization works were conducted on the hybrid energy harvester to evaluate the performance of the HEH system in powering the wireless sensor node. Figures 5.24 and 5.25 show the power curves of the HEH system at fixed solar irradiance of 380 and 1010 lux, respectively, for varying thermal differences in the range of 5 to 10 K. Conversely, the HEH system is also subjected to fixed thermal differences of 5 and 10 K, as shown in Figures 5.26 and 5.27, respectively, for various solar irradiances between 380 and 1010 lux.

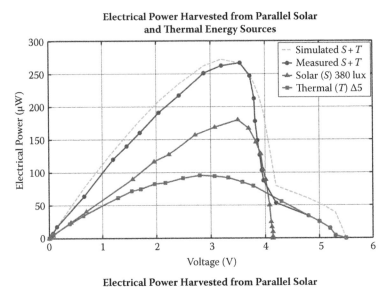

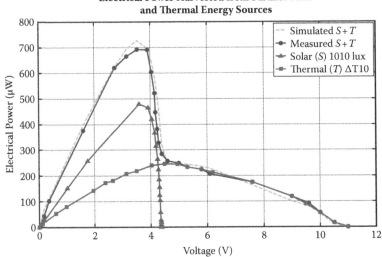

FIGURE 5.23
Compare experimental harvested power with simulated power under the least and most power harvesting conditions (top) 380 lux and $\Delta T = 5°C$ and (bottom) 1010 lux and $\Delta T = 10°C$, respectively.

Referring to Figures 5.24 and 5.25, it can be seen from the P-R curves (Thermal $\Delta T5$ - Thermal $\Delta T10$) that the MPPs of the stand-alone thermal energy harvester are fixed at its internal resistance of 82 kΩ under temperature differences in the range of 5 to 10 K. When the thermal energy harvester is parallel with the solar panel under weak illumination of 380 lux and strong illumination of 1010 lux, it can be observed from the P-R curves [Measured S+T ($\Delta T5$)

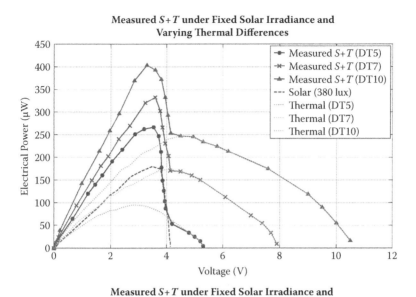

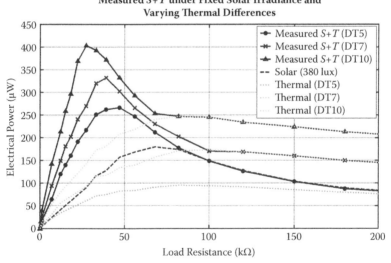

FIGURE 5.24

P-V and P-R curves of an HEH system at fixed solar irradiance of 380 lux (3 W/m^2) and different thermal differences of 5 to 10 K.

to Measured S+T (ΔT10)] shown in Figures 5.24 and 5.25 that the MPPs are no longer fixed, but vary with the combined internal impedance of the solar panel and the thermal energy harvester in parallel. This is the resulting effect of the impedance mismatch between the two energy sources of the hybrid energy harvester. Although there is an impedance mismatch issue in the proposed HEH system, it is still possible to combine the two energy sources together without dedicating each individual energy source with a power

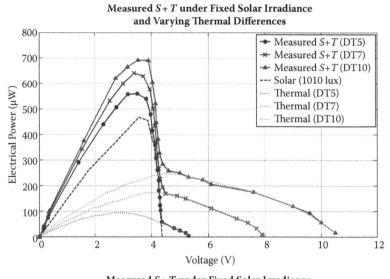

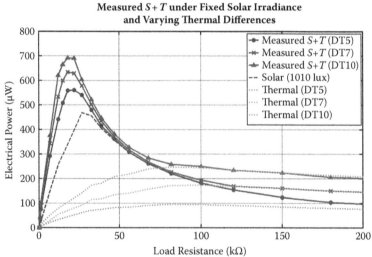

FIGURE 5.25

P-V and P-R curves of an HEH system at fixed solar irradiance of 1010 lux (3 W/m^2) and different thermal differences of 5 to 10 K.

converter to perform MPPT as proposed by Chulsung et al. [35]. Referring to Figures 5.24 and 5.25, the P-V curves [Measured S+T (ΔT5) to Measured S+T (ΔT10)] show that all the MPPs of the hybrid energy harvester are fixed at around its output voltage of 3.6 V.

Likewise, for an illumination level of 380 lux and above as a common indoor lighting condition, the hybrid energy harvester tested under fixed thermal difference of 5 and 10 K, as shown in Figures 5.26 and 5.27, show that all of the P-V power curves [Measured S+T (380 lux) to Measured S+T (1010 lux)] peak

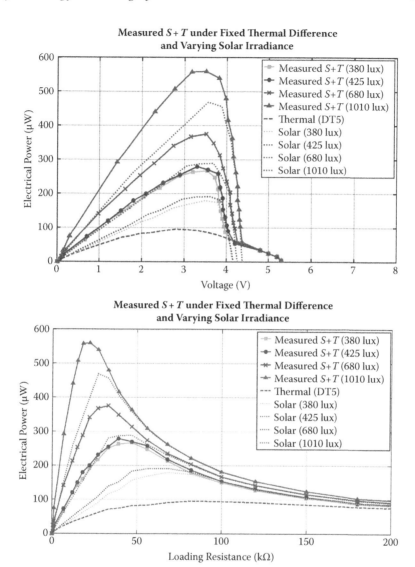

FIGURE 5.26

P-V and P-R curves of an HEH system at fixed thermal differences of $\triangle T = 5$ K and varying solar irradiances of 380 to 1010 lux.

around 3.6 V. Reviewing all the power curves shown in Figures 5.24 to 5.27, it can be observed that all the MPPs of the P-V curves tend to cluster around a fixed voltage of 3.6 V, whereas the MPPs of the P-R curves are scattered in the range of 20 to 50 kΩ. Hence, by deliberately setting the terminal voltage of the hybrid energy harvester to a value in the peak power range ($V_{R_L,mppt} = 3.6$ V), it is possible to extract maximum output power from the hybrid energy harvester with a simple and ultralow-power control circuit to place the panel

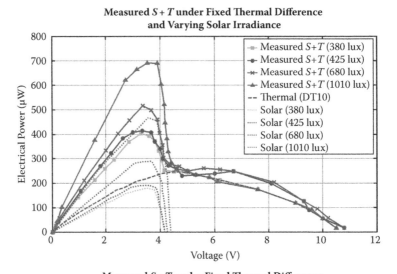

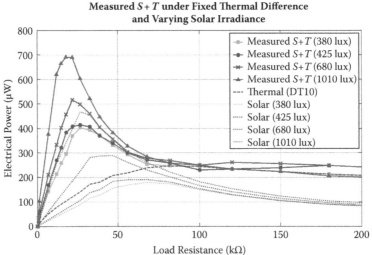

FIGURE 5.27
P-V and P-R curves of an HEH system at fixed thermal differences of $\Delta T = 10$ K and varying solar irradiances of 380 to 1010 lux.

at its MPPs, rather than using those energy-hungry tracking techniques, such as P&O and IncCond, which require high computational power and cost.

5.3.4.2 Design and Implementation of an Ultralow-Power Management Circuit

The schematic diagram of a self-autonomous indoor wireless sensor node powered by the proposed HEH system and its ultralow-power power management circuit is illustrated in Figure 5.28. Referring to Figure 5.28, the designed power management circuitry with fixed voltage reference MPPT

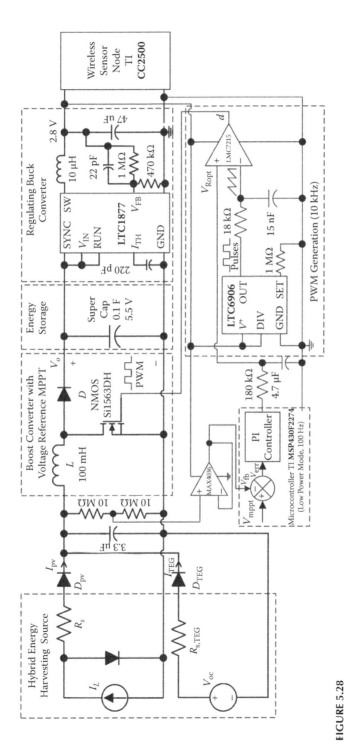

FIGURE 5.28
Functional block diagram of an HEH system.

approach essentially consists of three main building blocks: (1) a boost converter with MPP tracker and its control and PWM generation circuit that manipulates the operating point of the HEH scheme to keep harvesting power at near-MPPs; (2) an energy storage element (i.e., supercapacitor) to buffer the energy transfer between the source and the load; and (3) a regulating buck converter to provide constant voltage to the wireless sensor node and other electronic circuitries.

With reference to Figure 5.28, the operation of the boost converter based on the fixed voltage reference approach is given as follows: The MPPT voltage reference signal V_{mppt} of 3.6 V is compared with the feedback voltage signal V_{fb} from the output of the hybrid energy harvester. The resultant voltage error signal V_{err} is fed into a PI controller to generate a low-frequency PWM control signal, about 100 Hz, from a Texas Instruments microcontroller (TI MSP430F2274). In order to miniaturize the HEH system by using smaller passive components, the low-frequency PWM control signal generated from the reduced clock speed microcontroller is transformed to a higher switching frequency of 10 kHz. This is achieved by designing an ultralow-power PWM generation circuit made up of a micropower resistor set oscillator (LTC6906) used for sawtooth generation and a micropower, rail-to-rail complementary metal-oxide semiconductor (CMOS) comparator (LMC7215). The low-frequency PWM signal, which represents the MPPT voltage reference, is compared with the sawtooth signal to generate the high-frequency PWM gating signal to control the boost converter.

For an indoor environment, the ambient energy sources, such as a solar and thermal gradient, are not always available at all times and at a steady level, so there is a need to incorporate an energy storage device (i.e., supercapacitor) in the HEH system to store the excessive energy harvested from the solar panel or thermal energy harvester to buffer the indoor wireless sensor node for those times when energy sources are unavailable. Moreover, by drawing power simultaneously from both solar and thermal energy sources, the throughput power of the HEH system is increased, which can enhance the performance of the indoor wireless sensor node. A supercapacitor is employed in this work because it has superior characteristics over batteries as described by Simjee and Chou [34]. These characteristics include numerous full-charge cycles (more than half a million charge cycles), long lifetime (10 to 20 years operational lifetime), and high power density (an order of magnitude higher continuous current than a battery). Unlike the discrete capacitors, which have very small capacitance values in the picofarad-to-microfarad range, the supercapacitor has very large capacitance value farads in a range suitable for energy storage purposes.

Last, the switched-mode voltage regulator (LTC1877) obtained from Linear Technology is inserted after the supercapacitor to provide a constant operating voltage of 2.8 V_{DC} to the wireless sensor node and other electronic circuitries. The efficiency of the regulating buck converter was experimentally tested to be around 80% to 90%, consuming an operating current of 12 μA. In this work, the operation of the wireless sensor node deployed in an application field comprised of (1) sensing some external analog signals from sensory devices (such as temperature, humidity, etc.) and (2) communicating and relaying the sensed information to the gateway node every 5 s. On receiving the data at the base station, the collected data is then postprocessed into usable information for any follow-up action.

5.3.5 Experimental Results

The near-optimal HEH wireless sensor node has been successfully implemented into a hardware prototype for laboratory testing. Several experimental tests have been conducted to analyze the performance of the HEH system and its simple and ultralow-power fixed reference voltage MPPT scheme in powering the connected load consisting of the supercapacitor; the sensing, control, and PWM generation circuitries; and the wireless sensor node.

5.3.5.1 Performance of a Parallel HEH Configuration

As mentioned, when the solar and thermal energy sources of different characteristics are combined, it is bound to have impedance mismatch among the integrated energy sources. As such, the performance of the parallel hybrid energy harvester, which contains the combined characteristics of the solar panel as well as the thermal energy harvester, was investigated. With reference to Figures 5.24 to 5.27, it is illustrated that the fixed reference voltage method is able to operate the hybrid energy harvester near its MPPs for different light intensities and temperature differences, but at the expense of some percentage of power loss in the harvested power. It is thus important to examine the significance of these power differences between the actual harvested power $P_{HEH,actual}$ with respect to the MPPs $P_{HEH,mppt}$ of the hybrid energy harvester as recorded in Figure 5.29.

Considering an extreme operating condition, which is at low light illumination of 380 lux and small temperature difference of 5 K as seen in Figure 5.29a, for example, the power harvested at the fixed reference MPPT voltage of 3.6 V, $P_{HEH,actual}$, and the maximum obtainable power $P_{HEH,mppt}$ of the hybrid energy harvester is 252 µW and 260 µW, respectively. The power difference is only 8 µW, which is about 3% of its harvested power as shown in Figure 5.29b. The power loss of 8 µW is due to the impedance mismatch issue between the solar panel and thermal energy harvester when they are connected directly without the use of separate power converters.

Similarly, the power differences between the actual harvested power $P_{HEH,actual}$ with respect to the MPPs, $P_{HEH,mppt}$, for all the other operating conditions range between 8 and 35 µW (see Figure 5.29a), which are 3% to 6% of the harvested power (see Figure 5.29b). Although the proposed HEH system (i.e., hybrid energy harvester) and its power management unit would incur power loss (i.e., 8 to 35 µW) in the overall harvested power, this power loss is small and marginal compared to those MPPT techniques that require high computational power and cost to fulfill their objective of precise and accurate MPP tracking. It is thus justifiable to utilize the simple

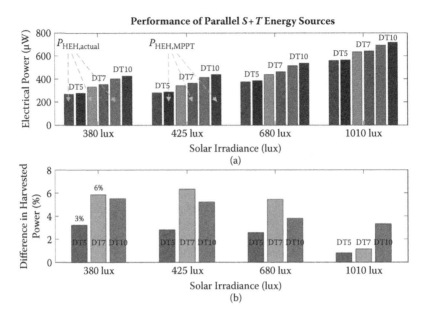

FIGURE 5.29
Performance of an HEH system in parallel configuration.

and ultralow-power fixed reference voltage method for the hybrid energy harvester.

5.3.5.2 Power Conversion Efficiency of the HEH System

The power conversion efficiency of the HEH system is another important investigation being carried out. Other than the regulating buck converter, there are two main contributors of power losses to the HEH system; the boost converter itself, which acts as an MPP tracker, and its associated sensing, control, and PWM generation circuits. The efficiency of the boost converter η_{conv} can be expressed as a function of its output load power P_{load} over its input DC power P_{dc} under varying solar irradiance, temperature differences ΔT, and loading R_L conditions. Take, for example, at lux illumination, temperature difference and output load resistance of 380 lux, 5 K, and 68 kΩ, respectively, the efficiency of the boost converter is given by

$$\eta_{conv} = \frac{P_{out}}{P_{in}} * 100\% = \frac{V_{out}^2 / R_{load}}{V_{in} I_{in}} * 100\% \tag{5.12}$$

$$= \frac{4.95 V^2 / 68 \text{ k}\Omega}{3.6 V * 109 \text{ μA}} * 100\% = 91.8\%$$

and the calculated efficiency of 91.8% is plotted in Figure 5.30. For all other solar irradiances, temperature differences, and loading conditions, the efficiencies of the boost converter are calculated using Equation 5.12, and the

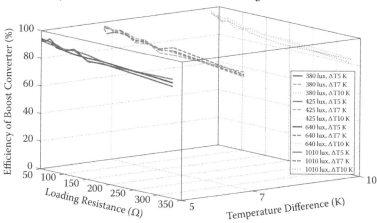

Efficiency of HEH Boost Converter with Fixed Voltage Reference-Based MPPT

FIGURE 5.30
Efficiency of HEH boost converter.

computed results are plotted in Figure 5.30. It can be observed from Figure 5.30 that the efficiency of the designed boost converter ranged between 80% and 94% over a range of load resistances of 50 to 330 kΩ. At heavy load condition, say 50 kΩ, which signifies the discharge state of the supercapacitor, it can be seen in Figure 5.30 that the efficiency of the boost converter was high, around 94%. This high-efficiency boost converter is very favourable and desirable to ensure optimal transfer of energy from the micropower sources of hundreds of microwatts or even less to the energy storage.

As the loadings lessen with the supercapacitor charge's increase, with reference to Figure 5.30, it can be seen that the efficiency of the converter decreases to around 82% at a load resistance of 300 kΩ. This decreasing efficiency trend is due to the power loss in the boost converter. Even though the efficiency at light load is lower, it is not as critical as the heavy load condition because the supercapacitor, by then, is already near the full charge state, and any surplus energy would not be stored. Another source of power loss in the HEH system is the power consumption of the associated sensing P_{sense} control, P_{ctrl}, and PWM generation $P_{generate}$ electronic circuits. Based on the voltage and current requirements of each individual component in the HEH system shown in Figure 5.28, the total power consumption of the electronic circuits can be calculated as follows:

$$P_{consumed} = P_{sense} + P_{ctrl} + P_{PWM\ generate}$$
$$= 2.7V * (3 + 15 + 32\ \mu A) = 135\ \mu W \qquad (5.13)$$

Once all the power losses in the HEH system are identified, including the power difference factor due to impedance mismatch between two paralleled energy sources and the power losses in the voltage-regulating and MPPT converters, the performance of the designed HEH system for enhanced per-formance in the indoor wireless sensor node was evaluated. In indoor appli-cations like hospitals and factories, say the ambient condition is as follows: solar irradiance of 1010 lux and temperature difference of 10 K, referring to Figures 5.17 and 5.20 with operating conditions of 1010 lux and 10 K, the maximum power obtained by summing the individual MPPs of the thermal energy harvester P_{TEG} and solar panel P_{pv} is 727 μW, and the actual har-vested power $P_{HEH,actual}$ measured from the two paralleled energy sources is 690 μW. The power difference between the calculated and measured pow-ers, due to impedance mismatch between two paralleled energy sources, is 35 μW, as shown in Figure 5.29. Taking into consideration both the power dif-ference and the power losses in the voltage-regulating and MPPT converters as shown in Figure 5.30, the net harvested power output to power the indoor wireless sensor node through the boost converter with efficiency of 90% is 621 μW. This harvested power from the HEH system is more than what is harvested by each individual energy source (i.e., ambient light of 432 μW or thermal energy source of 223 μW), hence the significance of the proposed HEH system is exhibited.

5.3.5.3 *Performance of the Designed HEH System for an Indoor Wireless Sensor Node*

In order to evaluate the designed HEH system for sustaining the operation of the wireless sensor node used in an indoor environment, the MPPT per-formance of the boost converter based on a fixed reference voltage scheme was investigated as shown in Figure 5.31. During the performance evalua-tion process, all the three main contributors of the power losses in the HEH system were also included in the power analysis. A 68-kΩ fixed resistor was used instead of a supercapacitor, which requires a long time to charge and discharge, to be the load so that the dynamic and steady-state responses of the MPP tracker could be examined for use in the power analysis. The load resistance represents the power consumed by the wireless sensor node for sensing and communicating operations as well as the power losses incurred by the electronic circuitries associated with the voltage-regulating and MPPT converters at a voltage level of 4.2 V.

Referring to Figure 5.31, it can be seen that from the initial start time t until the seventh second, there was only the presence of the heat source applied to the HEH system. The power harvested by the thermal energy harvester at a temperature difference ΔT of 10 K was about 190 μW at 3.6 V, which was less than the power required by the load. In order to sustain the operation of the wireless sensor node, an additional energy source from the artificial lighting with illumination of 380 lux was inserted at $t = 7$ s. It can be seen from Figure 5.31 that the output voltage of the boost converter V_o connected to the load increased from 3.6 to 5 V in around 5 s after which it settled down at 5 V. In the midst of that, the input voltage of the boost converter V_{in} surged and the HEH system shifted away from its MPP. Since a microcontroller-based MPP tracker with a closed-loop voltage feedback control was implemented

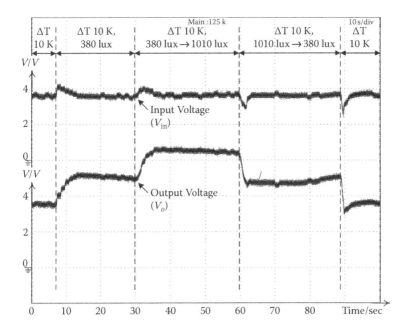

FIGURE 5.31
Performance of an HEH system.

in this section, the closed-loop MPP tracker manipulated the duty cycle of the boost converter to shift back V_{in} and maintain it always at the MPPT voltage ($V_{mppt} = 3.6$ V) of the HEH system. By doing so, the total power harvested by the HEH system increased from 190 to 367 μW at 5 V, which was more than the load power consumption.

To enhance the performance of the wireless sensor node, the solar panel was exposed to a stronger solar illumination of 1010 lux that can be found in indoor applications like hospitals and factories. Referring to Figure 5.31, for the time period of 30 to 60 s, there was even more electrical power harvested from the HEH system at a solar irradiance of 1010 lux and ΔT of 10 K for powering the indoor wireless sensor node. Referring to Figures 5.17 and 5.20 with operating conditions of 1010 lux and 10 K, the maximum power obtained by summing the individual MPPs of the thermal energy harvester and solar panel was 727 μW, and the actual harvested power measured from the two paralleled energy sources was 690 μW. The power difference between the calculated and measured powers, due to impedance mismatch between two paralleled energy sources, was 35 μW. Taking into consideration both the power difference and the power losses in the voltage-regulating and MPPT converters, the net harvested power output to the load through the boost converter with efficiency of 90% was 621 μW. This verified the harvested power experimentally obtained from Figure 5.31, where the output voltage V_o of the boost converter connected to the load resistance of 68 kΩ was around 6.5 V, and its harvested power was calculated to be 621 μW.

5.3.6 Summary

A near-optimal HEH system has been proposed for enhanced performance of an indoor wireless sensor node. Theoretical studies of individual subsystems as well as hybrid SEH and TEH systems were conducted and simulated to understand the characteristic of the HEH system, which were then verified by experimental results. In this section, the proposed HEH system using one power management circuit was successfully implemented into a hardware prototype for laboratory testing. Based on the power analysis, the efficiency of the power management unit with fixed reference voltage-based MPPT scheme was around 90% and its sensing, control, and PWM generation circuitries consumed around 135 μW. Experimental results showed that the HEH system can harvest an average electrical power of 621 μW from both energy sources at an average solar irradiance and thermal gradient of 1010 lux and 10 K, respectively, which is almost 3.3 times higher than the conventional single thermal energy harvesting method to enhance the performance of the indoor wireless sensor node.

6

Electrical Power Transfer with "No Wires"

In this chapter, the concept of electrical power transfer with "no wires" is introduced with a similar objective as energy harvesting, that is, to enable remote charging of low-power electronic devices. Wireless power transfer (WPT) is the process by which a system delivers electrical energy from a power source to the load without any connecting wires. Almost 100 years ago, before the advent of the electric grid, the genius inventor Nikola Tesla envisioned a future when huge towers would radiate energy directly into our homes for consumption [147]. But only over the past 5 years has there been a remarkable interest in researching commercially viable and safe methods of WPT, mainly due to the surge in the use of low-power electronic devices like laptops, netbooks, smart phones, wireless sensor nodes, and more that require regular charging and battery maintenance. It is obvious that WPT products can increase convenience and quality of life, but what is not so obvious are the environmental and economic benefits that this technology may have to offer. There is great potential in using WPT to directly power devices such as clocks and remote controllers, which would drastically reduce the 6 billion batteries being disposed of every year around the world that are a source of groundwater contamination and producers of toxic waste when burnt in incinerators. In other cases where wiring is too expensive, hazardous, or impossible, WPT may be the only enabling technology.

Presently, several WPT techniques are being pursued, which can be categorized in terms of their underlying power transfer mechanism to understand the implications for range, adaptation, and efficiency. Far-field WPT is one of the emerging techniques that uses propagating electromagnetic (EM) waves to transfer energy. This method has been successfully used to power radio-frequency identification (RFID) tags; which have no batteries and an operating range of about 10 m [148–149]. One of the drawbacks with far-field WPT approaches is the inherent trade-off between directionality and transmission efficiency. There are many examples of RF and microwave systems that use lasers or high-gain antennas to transfer power over kilometre distances at efficiencies of over 90% [150–151]. These systems suffer from the need for sophisticated tracking and alignment equipment to maintain an uninterruptible line of sight (point-to-point) connection in unstructured and dynamic environments. Alternatively, RF broadcast methods, which transmit power in an omnidirectional pattern, allow for power transfer anywhere in the coverage area. In this case, mobility is maintained, but end-to-end efficiency is lost since power density decreases with a $1/r^2$ dependence, resulting in received power levels many orders of magnitude less than what is transmitted [152]. In order to provide power comparable to a typical wall-mounted direct current (DC) power adapter, the system would violate RF safety regulations [153] or have to use a large number of transmitters, resulting in an impractical implementation. Therefore, far-field techniques are most suitable for very-low-power applications unless they are used in less-regulated environments, such as military endeavours or space exploration.

According to Low et al. [153], inductive coupling, which is nonradiative and near field, has been one of the leading candidates in achieving WPT at power levels ranging from several microwatts to several kilowatts. Its operating range is limited as power delivery and efficiency degrade rapidly with increasing distance between the transmitting and receiving units. Inductive coupling, which does not rely on propagating EM waves, operates at distances less than a wavelength of the signal being transmitted [154]. Applications include rechargeable toothbrushes and the recently proliferating "power" surfaces [155]. These techniques can be very efficient but are limited to transmission distances of about a centimetre. Alternatively, near-field RFID pushes the limit on distance by sacrificing efficiency. Near-field tags have a range of tens of centimetres, but only receive power in the microwatt range with 1% to 2% transmission efficiency [148]. Previously demonstrated magnetically coupled resonators used for WPT [156–158] have shown the potential to deliver power with more efficiency than far-field approaches and at longer ranges than traditional inductively coupled schemes. However, this prior work is limited to a fixed distance and orientation, with efficiency falling off rapidly when the receiver is moved away from its optimal operating point. This chapter explores both types of near-field WPT techniques for remote charging of low-power electronic devices : (1) inductively coupled WPT from power lines (see Section 6.1) and (2) strongly coupled WPT with magnetic resonances (see Section 6.2).

6.1 Inductively Coupled Power Transfer from Power Lines

Inductive coupling has considerably increased in the last 10 years because it permits the supply of power to electronic circuits remotely and also provides a means for exchange of data between two sensor nodes. Today in the industry, the most important part of inductive coupling applications is related to RFID [159]. RFID is used for contactless smart cards or to sort and locate travelling objects (e.g., postal parcels). In many buildings, door keys have been replaced by the use of EM badges. However, existing technologies that utilize near-field wireless power transmission such as RFID tags generally operate only over limited distances with very low efficiency. Other than RFID application, power transmission and remote sensing by inductive coupling are also widely used in the field of sensors. Yang et al. [160] explored the concept of distributed sensing along power lines using sensor nets. Power lines span long distances, and sensing from remote substations, as is currently done, provides poor resolution. Hence, distributed sensing along the length of the power line with intelligent sensors powered by inductive coupling would allow extraction of important local information from each sensor for a long period of time. Many possibilities of incorporating inductive coupling in medical applications as well are foreseen, for example, permanent implants, supervising various parameters like the insulin rate for diabetics or intraocular blood pressure, and excitation of a retina with defective photoreceptors [161].

Although many research works on inductive coupling power transfer (ICPT) have been reported in the literature, such as by Boys et al. [162] and Kurs et al. in [163], there are only a few research works on magnetic energy harvesting for wireless sensor nodes, like the working Yang et al. [160]. They have proposed the use of inductive coupling from high-power transmission and distribution lines to power the sensor nets for power grid monitoring. In this section, the proposed magnetic energy harvesting idea is different from what Yang et al. have suggested. The main objective of this research work is to harvest the stray magnetic energy generated by electrical power cables deployed around the residential, commercial, and industrial buildings. Through inductive coupling, the stray magnetic energy is harvested to power the wireless sensor nodes for condition-based maintenance of electrical energy systems, and equipment.

6.1.1 Magnetic Energy Harvester

The research work on magnetic energy harvesting via inductive coupling utilizes induction as the energy harvesting technology. It is based on the combination of the famous Ampere's law and Faraday's law of induction. Ampere's law describes the magnetic flux density of the stray magnetic energy source available for induction by the surge coil. Faraday's law of induction states the induced electromotive force V_{emf} in a surge coil is directly proportional to the time rate of change of magnetic flux ϕ through the winding loop. The induced voltage V_{emf} generated at the output of the surge coil is processed by a power management unit and stored in an energy storage device (i.e., capacitor). This stored energy is then used to power up the operation of a wireless sensor node.

In experimental tests, the characterization process of the magnetic energy harvester was divided into two parts: (1) the magnetic energy source (i.e., magnetic field containing the magnetic energy governed by Ampere's law) and (2) the magnetic energy harvester (i.e., toroid-based surge coil wound with many turns of wires N as described by Faraday's law of induction). The magnetic energy source is first characterized using the experimental setup shown in Figure 6.1. Since the magnetic flux density B along a current-carrying electrical power cable is a function of the current I flowing through

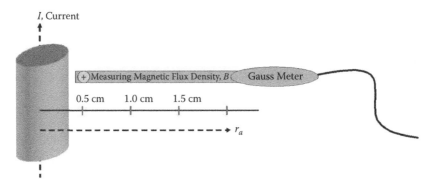

FIGURE 6.1
Characterization of a magnetic energy source based on Ampere's law.

the power cable and the radius distance r_a between the measurement point and the centre of the conductor, it is possible to determine the magnetic field lines that best describe the magnetic energy generated by the current flowing in the power line.

The second part of the characterization process is to determine the voltage induced by the toroid-based magnetic energy harvester, constructed by physically winding N turns of copper wires on a circular ring-shaped ferrite magnetic core as shown in Figure 6.2. When the current-carrying power cable is laid through the centre of the ferrite core as illustrated in Figure 6.3, magnetic field lines are generated. These magnetic field lines circulate around the ferrite core and its copper winding and an AC (alternating current) voltage is induced. The induced voltage is proportional to the rate of change in the number of flux lines enclosed by the loop per unit time and the number of winding turns N in the loop. In other words, the induced voltage is related

FIGURE 6.2
Top view of ferrite core windings.

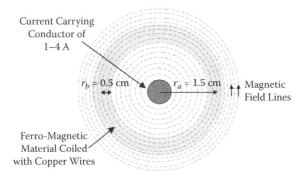

FIGURE 6.3
Top view of ferrite core with current carrying conductor.

to the magnetic field B, the loop area A, the winding number of turns N, and the frequency of the current f.

The experimental setup to measure the induced voltage V_{emf} [$= \omega NB$ $A \sin(\omega t)$ where the negative sign is due to Lenz's law] from the toroid-based magnetic energy harvester via the current-carrying power cable is shown in Figure 6.4. The current flowing in the circuit varies from 1 to 4 A by adjusting the voltage knob of the AC power supply. Due to the high current along the circuitry, the high-wattage resistor load bank is utilized. A summary of the measured and calculated induced emf (electromotive force) voltage for difference current flowing in the primary side power line is presented in Table 6.1.

It can be observed from Table 6.1 that as the current flowing in the mainstream power line increased from 1 to 4 A, the magnetic field B obtainable at 1.5 cm away from the centre of the conductor also increased from 0.02 to 0.08 T. For that reason, the induced voltage generated at the output of the toroid magnetic energy harvester was increased. During the characterization process, the r_a distance of 1.5 cm was set as the reference point based on the practical considerations of the physical diameter of the power cables and the space taken by 500 turns of copper winding.

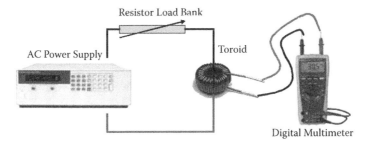

FIGURE 6.4
Characterization of a magnetic energy harvester based on Faraday's law.

TABLE 6.1

Measured and Calculated-Induced EMF Voltage for Different Current Flowing in the Power Line

I, Calculated Flowing in Power Line	μ_r	ω $2\pi f$ f at 50Hz	N	B $\mu_0\mu_r I/2\pi r_a$ r_a at 1.5cm	Area (πr_b^2) $r_b \sim 0.5$cm	Current V_{emf} (V_{rms})	Measured V_{emf} (V_{rms})
				Measured and Calculated Induced EMF Voltage $V_{emf} = \omega NBA\sin(\omega t)$			
4A	1500	$100\,\pi$	500	0.08 T	$2.5 \times 10^{-5}\pi$	0.987	1.025
3A	1500	$100\,\pi$	500	0.06 T	$2.5 \times 10^{-5}\pi$	0.740	0.748
2A	1500	$100\,\pi$	500	0.04 T	$2.5 \times 10^{-5}\pi$	0.493	0.449
1A	1500	$100\,\pi$	500	0.02 T	$2.5 \times 10^{-5}\pi$	0.247	0.194

6.1.1.1 Performance of the Magnetic Energy Harvester

To study how the magnetic energy harvester performed under various operating conditions, experiments were carried out on the designed magnetic energy harvester. Referring to Table 6.1, the open-circuit voltage of the harvester, which consists of one toroid coil, was quite low, ranging from 0.2 to 1 V. In order to achieve a higher output voltage, three sets of ferrite cores were connected in series. The improved version of the magnetic energy harvester was connected to different loading resistances, and the source current flowing in the power line was varied between 1 to 4 A. This was to find out the performance of the harvester for various input and output operating conditions. The data collected is plotted into the current versus voltage (I-V) curve and power curve as shown in Figures 6.5 and 6.6, respectively.

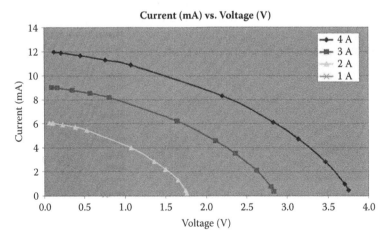

FIGURE 6.5
Voltage versus current curve for various input currents.

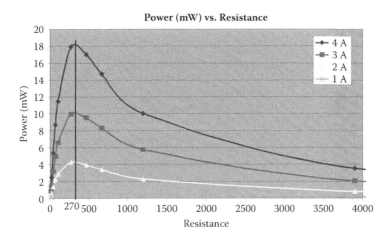

FIGURE 6.6
Electrical power harvested over a range of load resistances for different input currents.

From the I-V curve shown in Figure 6.5, it can be seen that the obtainable open-circuit voltage for different currents flowing in the power line increased by around three times, ranging from 0.7 to 3.5 V. Although the output voltage of the harvester had already been increased by series connecting three ferrite coils, the induced voltage at some operating points, especially when the magnetic field was weak due to low current flows in the AC power line, was considerably low. As such, the magnetic energy harvester may not be able to drive the electronic output load. This low output voltage generated from the magnetic energy harvester would pose a challenge for the design of the power management circuit. Another analysis carried out was on the power curve shown in Figure 6.6, where maximum power was attainable at a load resistance of 270 Ω. Referring to Figure 6.6, with the source current of 1 to 4 A flowing in the AC power line, the maximum electrical power available for harvesting ranged from 1 to 18 mW. The challenge here was that when the source current was low, say 1 A, the radiated magnetic field became weak, so the maximum power available for harvesting dropped tremendously to around 1 mW or so and may not be sufficient to power the RF transmitter load continuously. Hence, a power management circuit designed to address the low-voltage and low-power challenges of the magnetic energy harvester has been proposed.

6.1.2 Power Management Circuit

Based on the analysis and characterization performed on the designed magnetic energy harvester, the concept of harvesting stray magnetic energy via ICPT was found to be a viable solution for powering the low-power wireless sensor nodes. The block diagram in Figure 6.7 illustrates the energy harvesting scheme and its application for wireless sensor nodes. Since the voltage source

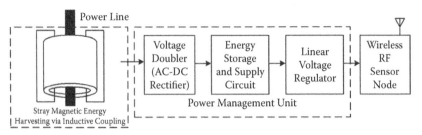

FIGURE 6.7
Block diagram of energy harvesting and a wireless RF transmitter system.

is inherently AC from the power supply along the power line, the induced voltage V_{emf} would appear as an alternative voltage source to the connected load. However, the wireless sensor node (i.e., AM RF transmitter) requires a DC source to operate; therefore, the induced voltage must be rectified to DC and regulated prior to powering up the device. This is achieved by using a voltage doubler instead of a standard diode-based full-wave rectifier, which is capable of rectifying and amplifying the low AC voltage to a higher DC voltage.

Referring to the power curve shown in Figure 6.6, the amount of power that is generated across the designed ferrite core wound with copper wires is in the few milliwatt range. With the limited power generation level, it is not feasible for the magnetic energy harvester to power the wireless RF transmitter continuously. To overcome that, an effective energy storage and supply circuit discussed by Tan et al. [166] was designed and inserted between the energy source and the wireless load. This ensured that the electrical energy was stored in the capacitor and the energy stored was sufficient to sustain the operation of several RF transmissions. When the energy level of the storage capacitor in the power management unit was sufficient for operations, the RF transmitter would then start to transmit digital-encoded information to the RF receiver located some distance away. The amount of energy consumed by the transmitter was dependent on the number of 12-bit digital-encoded data to be transmitted.

The design specifications of the experiments to be carried out in the research work were defined according to the practical field condition. The specifications were as follows: (1) source frequency of 50 Hz, which is the operating frequency in the Singapore context; (2) electrical current flowing in the mainstream power line set at 4 A; and (3) 500 turns in each winding. The advantage of the magnetic energy harvester is that it provides flexibility in the design parameters, that is, N, ω, B, involved, which can be designed accordingly to suit different operating conditions of the sensor node in certain specific applications. The experimental setup of the entire magnetic energy harvesting system, which consists of the stray magnetic energy harvester, power management unit, and wireless RF transmitter, is shown in Figure 6.8.

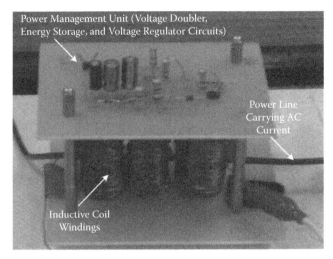

FIGURE 6.8
Photograph of the magnetic energy harvester system powering the wireless RF transmitter.

Based on the electrical power requirements of the wireless RF transmitter and the electrical characteristic of the magnetic energy harvester (i.e., N, ω, and B), a magnetic energy harvesting prototype has been designed and successfully implemented. The schematic drawing of the prototype of the magnetic energy harvesting system is shown in Figure 6.9. The prototype was designed to be similar to a transformer secondary winding capable of inducing AC voltage from the current-carrying conductor single-turn primary winding to power up a wireless RF transmitter. The induced V_{emf} was first rectified to DC voltage using a voltage doubler. Once the voltage from the secondary coil of the transformer was rectified, the electric current would flow to charge the electrolytic storage capacitor C1 of the power storage and supply circuit. The Q1 and Q2 metal-oxide-semiconductor field-effect transistors (MOSFETs) residing in the power storage and supply system acted like a control switch that would initiate an on or off signal to the storage capacitor to release the stored energy. Initially, both Q1 and Q2 were off, so the ground lines of the linear regulator (MAX666) and the RF AM transmitter (RTFQ1-433) were disconnected from C1. As C1 charged beyond the preset voltage of around 6.8 V (the preset voltage level was determined according to the induced voltage V_{emf} in Figure 6.9 and the voltage level was preset by the zener diode Z1 as 6.2 V and the gate-source junction of Q1 as 0.6 V), the control switch Q1 turned on. The moment when Q1 was on, there was a voltage drop across R2 that was higher than the threshold gate-source voltage $V_{gs(th)}$ of Q2 in order to activate the control switch Q2. Once Q2 was activated, Q1 was latched. This connected the ground lines of MAX666 and AM-RTFQ1-433 with C1, allowing C1 to discharge through the circuitry. MAX666 acted like a low power series linear regulator, which produced a stable +3.3 V for the serial ID encoder (HT12E) and the RF AM transmitter (AM-RTFQ1-433)

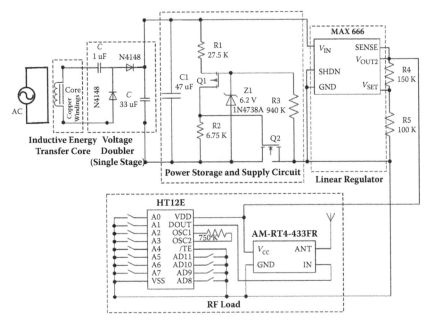

FIGURE 6.9
Schematic drawing of the magnetic energy harvesting system.

throughout the discharge of C1. When the voltage across C1 dropped below around 4 V, the voltage drop across R3 caused Q1 to turn off and hence in turn deactivated Q2 from the latched stage. When this happened, the ground lines of MAX666 and AM-RTFQ1-433 were disconnected from C1, and the discharge of C1 was stopped. As the secondary coil of the transformer continued to induce V_{emf} to the voltage doubler and increased the voltage on C1, the cycle started again.

6.1.3 Experimental Results

An experimental platform, which consisted of a 220/230 V_{ac} power supply connected to a bank of load resistances of 60 Ω, was set up as shown in Figure 6.10 to emulate the electrical current of 1 to 4 A flowing in the power line [167]. Since the primary side power line was AC, the induced emf would be AC voltage. This experimental setup was used as a testing platform to evaluate the performance of the magnetic energy harvesting system.

Figure 6.11 shows the waveforms of the induced AC voltage V_{emf} output of the stray magnetic energy harvester and the output DC voltage of voltage doubler circuit. It can be observed that the induced voltage signal is a distorted sinusoidal wave rather than a smooth sinusoidal wave. The reason for this phenomenon may be due to the magnetic hysteresis effect and the magnetic saturation of the toroid core.

FIGURE 6.10
Experimental testing platform for the magnetic energy harvester.

Once the induced AC voltage V_{emf} of the stray magnetic energy harvester was input to the voltage doubler, the voltage doubler circuit output a DC and doubled voltage. By doing so, the design of the power management would be simpler. The voltage from the secondary coil of the transformer was rectified in the voltage doubler, and then electric charge was accumulated on the electrolytic storage capacitor C1 of the power storage and supply circuit. The charging and discharging voltages of C1 were 4 and 6.72 V, respectively (as shown in Figure 6.12). The amount of electrical energy stored in the electrolytic capacitor C1 with capacitance value of 47 µF was calculated to be 685 µJ.

For each 12-bit digital data, the time taken for one transmission was 20 ms, that is, 10 ms of active time and 10 ms of idle time. During the active transmission time, the supply voltage and current of 3.3 V and 4 mA, respectively, were consumed by the RF transmitter load. As for the remaining time of 10 ms, the RF transmitter load was operating in idle mode, which means that a very minimal amount of energy would be consumed, so it is reasonable to exclude the power being consumed by the RF transmitter load during the

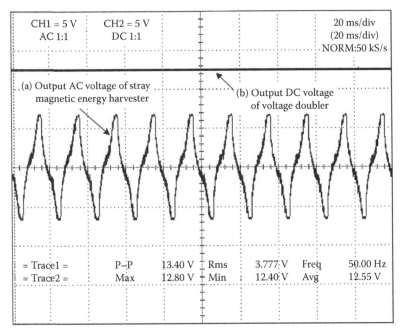

FIGURE 6.11
Waveforms of (a) output AC voltage of stray magnetic energy harvester and (b) output DC voltage of voltage doubler.

idle time. By calculation, the average power and hence the energy consumed by the RF transmitter load for one digital encoded data transmission were 13.2 mW and 132 μJ, respectively. Using the harvested stray magnetic energy in the power lines via inductive coupling, the experimental results shown in Figure 6.12 verified that the RF transmitter was able to successfully transmit more than 10 digital-encoded data to the receiver remotely. This was verified by the number of digitally encoded data packets received at the RF receiver side.

With the experimental platform, the power harvested and consumed by the source and the load, respectively, can be measured, and this is illustrated in Figure 6.13. The line diagram shows the flow of power from the harvested power through inductive coupling to the power management circuit, which consisted of the voltage doubler circuit, energy storage and supply circuit, and regulator circuit, and then to the RF transmitter load. It can be seen that the power generated by the source was not sufficient to power the load directly and continuously; therefore, an energy storage element with a supply control feature was placed between the source and the load. By doing so, the operation of the RF transmitter was no longer kept running; instead, it was more like an intermittent type of operation. Although the amount of energy that the stray magnetic energy harvester can harness was relatively small

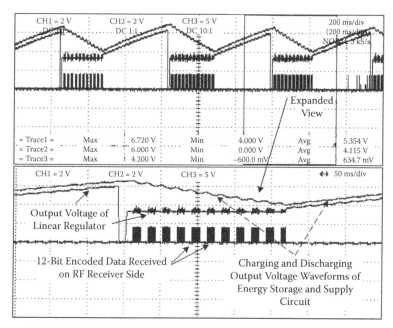

FIGURE 6.12
Waveforms collected at the RF receiver side to display the number of 12-bit encoded data packets received using the harvested energy.

as compared to other energy harvesting sources (i.e., solar and wind), the small amount of energy of 685 μJ was sufficient to power its RF transmitter load to transmit several digital-encoded data in wireless transmission. The experimental results demonstrating the successful RF transmission using the harvested energy are shown in Figure 6.12 . This implies that the magnetic energy harvester was able to meet its objective.

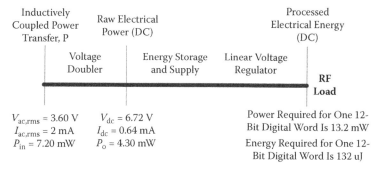

FIGURE 6.13
Line diagram of the power flow in the system.

6.1.4 Summary

In this research work, the inductive coupling concept was utilized for wireless power harvesting from power lines. Based on Faraday's law of induction and Ampere's law, a toroid-based surge coil wound with 500 turns of wires was designed as a magnetic energy harvester. The magnetic energy radiated by a 50 Hz, 4 A, 230 V_{ac} current-carrying power line/cable was harvested by the designed magnetic energy harvester via inductive coupling to power a wireless sensor node. The prototype of the inductive energy transfer system was designed and developed to convert harvested AC voltage into DC voltage, which then charged the storage capacitor until the preset energy storage level. After this, the stored energy was released to the linear regulator to provide a constant 3.3 V to the RF AM transmitter which required 132 µJ of energy for communication. Finally, the self-powered wireless RF transmitter working prototype was capable of transmitting 10 packets of 12 digital bits information over a range of up to 70 m in an open field with line of sight.

6.2 Wireless Power Transfer (WPT) via Strongly Coupled Magnetic Resonances

In recent years, mobile gadgets such as laptops, iPods, iPads, mobile phones, and digital cameras have become common in the consumer world. These electronic devices and gadgets are typically powered by either the AC main supply through power cords or solely alkaline/rechargeable batteries that need to be replaced or recharged regularly. People are getting tired of the mess, as illustrated in Figure 6.14, created by power cords and chargers of multiple electronic devices and the trouble of not being able to locate these accessories when there is a need to recharge. When people forget to replace the alkaline batteries in some electronic products, it damages the product due to leakage and chemical reactions, which may lead to the disposal of the whole product. In other cases, extensive electrical wiring laid out from the main power supply to the user is too expensive and hazardous. Apart from the technical issues, people are also concerned about the environmental and economical impacts arising from the disposal of batteries.

Because of these concerns, there is an emergence of interest in the WPT technology to deal with these challenges. WPT is neither a crazy idea nor a very new concept. If WPT can be realized in practical applications, the outcome of WPT technology could potentially revolutionize the way electrical energy is used for consumer electronic devices and portable gadgets and thereby reduce the dependence on plug-in power supply and disposable batteries to achieve the ultimate goals of *wireless* and *batteryless* electronic products. There are several wireless charger-based products [168] —Powermat, WildCharge, Fulton Innovation's eCoupled, Powercast, and so on—already available on the market. These radiative and nonradiative WPT products are mostly for

FIGURE 6.14
Tangled mess of power cords and chargers.

near-field applications with short distance of a few centimetres. Some application examples include the commonly used electric toothbrush chargers that transfer power to the brush handle through magnetic induction, a wireless inductive charger for the iPhone [169], and more. In order to elongate the wireless energy transmission distance, WPT with magnetic resonance was first reported by Kurs et al. [163], and subsequently the concept has been implemented and demonstrated by companies like Sony [170]. The prototype system developed by Sony is able to reach up to 60% power efficiency with maximum power delivery at 60 W for a range of 50 cm. However, as the distance between the coils increases slightly, the efficiency and power throughput of the WPT system drops tremendously. Hence, there is still lots of room for improvement in the WPT technology used in products. What is lacking in these products is the combination and extension of WPT into a system to be incorporated in daily life.

In this section, the research focus is on the analysis, design, and implementation of WPT resonator coils with optimal efficiency and form factor to charge consumer electronic devices and portable gadgets in a connected manner. The research work was divided into two parts: (1) to explore different ways to improve the efficiency of the WPT technique operating in a strongly coupled regime and (2) to network the designed high-efficiency WPT resonator coils together to achieve the ultimate goals of wireless and batteryless electronic products. Several design factors, which include coil size, coil structure, and configuration and distance between coils, that could enhance the efficiency of the WPT system were explored and experimentally verified. The optimized resonant coils were then networked together to receive and relay electrical power from one resonant coil to another nearby resonator coil. By doing so, the WPT distance between each gadget can be reduced and the form factor of the resonator coil can become smaller in size to fit into the mobile gadget. It is thus more viable to utilize WPT technology as a self-sustaining portable energy source for the mobile gadgets.

6.2.1 Concept Principles of WPT with Magnetic Resonance

The concept of WPT is very much similar to an air-core transformer where no ferromagnetic material is inserted between the primary and secondary sides of the transformer. The source power of alternating frequency is connected to the primary coil, setting up a constantly changing magnetic field, which induces an electromotive force in the secondary coil according to Faraday's law of induction. This electromotive force then drives an induced current in the secondary coil. Efficiency of a transformer is maximized by having the secondary coil tightly wound around the primary in order to maximize the coupling between the coils. As the magnetic field strength, which is given by the Biot-Savart equation as

$$|dB| = \frac{\mu_o I dl}{4\pi r^2} \tag{6.1}$$

decreases with the square of the distance r separating the secondary coil from the primary coil, the magnetic coupling between the coils is greatly reduced, hence the efficiency of the transformer or the WPT system. To strengthen the magnetic coupling, magnetic resonance for WPT has been discussed by Kurs et al. [163] such that both the primary (transmitting) and secondary (receiving) coils are tuned to the same resonant frequency, making use of magnetic resonance to compensate for the weakened coupling due to the increase in separation distance. Magnetic resonance occurs when magnetic waves of certain frequencies are absorbed by an object, causing that object to resonate. The frequency absorbed by an object that causes that object to resonate is dependent on many variables, including molecular structure, shape, and size/length of the object. In an experiment conducted by Jorgensen and Culberson [171], the authors had a primary coil emitting evanescent, magnetic waves that a secondary coil absorbed. Evanescent waves are different from ordinary waves because evanescent waves oscillate in time but diminish over distance [171]. Assuming the coils are of the same size, shape, and mass, at a certain frequency (the resonant frequency) the primary coil will resonate and cause the secondary coil to resonate as well. An overview of the WPT system is shown in Figure 6.15.

According to Kurs et al. [163], intuitively, two resonant objects of the same resonant frequency tend to exchange energy efficiently, while dissipating relatively little energy in extraneous off-resonant objects. In systems of coupled resonances, there is often a general "strongly coupled" regime of operation. If one can operate in that regime in a given system, the energy transfer is expected to be very efficient. Midrange power transfer implemented in this

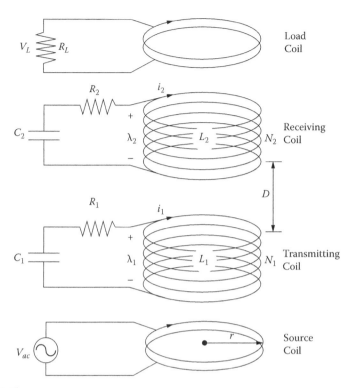

FIGURE 6.15
Overview of a WPT system.

way can be nearly omnidirectional and efficient, irrespective of the geometry of the surrounding space, with low interference and losses into environmental objects. Based on this magnetic resonance concept, the transmitting and receiving coils of the WPT system were made of conducting loops with a capacitor attached to the ends. For this inductor-capacitor (LC) configuration, the resonant frequency of the circuit is determined by the equation

$$\omega = \frac{1}{\sqrt{LC}} \tag{6.2}$$

where ω is the resonant angular frequency, L is the inductance of the coil, and C is the capacitance value of the external capacitor connected in parallel to the coil. At resonant frequency ω, the resistor-inductor-capacitor (RLC) circuit acts like an oscillator/resonator that transfers energy between the capacitor C and the inductor L, where the collapsing magnetic field of the inductor generates an electric current in its windings, which in turn charges the capacitor, and then the discharging capacitor provides an electric current that energizes the magnetic field in the inductor. This energy conversion process repeats itself periodically. When the RLC circuit is set to operate at its resonant frequency, the complex part of the circuit's impedance, which is expressed as

$$Z = R + j\left(\omega L - \frac{1}{\omega C}\right)$$ (6.3)

becomes zero, so the impedance of the electrical circuit is at its minimum, $Z = R$. As such, it is desirable to design the RLC circuit to operate at its resonant frequency, as expressed by Equation 6.2, by selecting the appropriate L and C based on the construction of the coils and the capacitors.

In this WPT research work, the objective was to explore different ways to improve the efficiency of the WPT technique operating in the strongly coupled regime. The efficiency η of the WPT system with magnetic resonance in accordance to the coupled-mode theory is a function of the coupling-to-loss ratio κ/Γ, where κ is the coupling coefficient of the coils and Γ is the intrinsic loss rate, and it is expressed [163] as

$$\eta = \frac{\frac{\Gamma_W}{\Gamma_D}\frac{\kappa^2}{\Gamma_S\Gamma_D}}{\left[\left(1 + \frac{\Gamma_W}{\Gamma_D}\right)\frac{\kappa^2}{\Gamma_S\Gamma_D}\right] + \left[\left(1 + \frac{\Gamma_W}{\Gamma_D}\right)^2\right]}$$ (6.4)

where the source and device are identified by subscript S and D, respectively, and an external load (subscript W) acts as a circuit resistance to the connected device. Referring to [163], Kurs et al. state that the efficiency of the WPT system is maximized when $\Gamma_W/\Gamma_D = [1 + (\kappa^2/\Gamma_S\Gamma_D)]^{1/2}$. In addition, for two identical coils, $\Gamma_S = \Gamma_D = \Gamma$, the efficiency of the WPT system can thus be expressed as

$$\eta = \frac{\frac{\kappa^2}{\Gamma^2}\sqrt{1 + \frac{\kappa^2}{\Gamma^2}}}{\left[\left(1 + \sqrt{1 + \frac{\kappa^2}{\Gamma^2}}\right)\frac{\kappa^2}{\Gamma^2}\right] + \left[\left(1 + \sqrt{1 + \frac{\kappa^2}{\Gamma^2}}\right)^2\right]}$$ (6.5)

For the WPT system to be efficient, the magnetic fields of the coils must be strongly coupled such that the coupling coefficient of the coils κ is high and the intrinsic loss rate Γ is low. If that is the case, $\kappa/\Gamma \gg 1$, and the efficiency of the WPT system with magnetic resonance expressed in Equation 6.5 can be further deduced as follows:

$$\eta = \frac{\frac{\kappa}{\Gamma}}{\frac{\kappa}{\Gamma} + 1} \approx 1$$ (6.6)

Hence, to maximize the efficiency of the WPT system, one important factor to be considered is the coupling coefficient κ of the coils [163], which is given by

$$\kappa = \frac{\omega M}{2\sqrt{L_1 L_2}}$$ (6.7)

where L_1 and L_2 are the inductances of the primary and secondary coils, respectively. M is the mutual inductance of the coils, derived from Biot-Savart's law using simple approximations [172], which is expressed as

$$M = \frac{\mu_o \pi N^2 r^2 r_b^2}{2 \left(r^2 + D^2\right)^{1.5}} \tag{6.8}$$

where $\mu_o = 4\pi \times 10^{-7}$ NA^{-2} is the permittivity of free space; r and r_b are the primary and secondary coil radii, respectively. N is the number of turns in the coil, and D is the distance between coils. Another consideration factor is the intrinsic loss rate Γ of the WPT system, which is discussed [163] with the following equations, representing the ohmic or absorption loss R_{ohmic} and radiation loss $R_{radiative}$. The intrinsic loss rate is defined by the following equation:

$$\Gamma = \frac{R_{ohmic} + R_{radiative}}{2L} \tag{6.9}$$

For a coil with N turns, radius r, and height h and made of an electrically conducting wire with conductivity of σ, the ohmic resistance R_{ohmic} is expressed as

$$R_{ohmic} = \frac{l}{4\pi a}\sqrt{\frac{\mu_o \omega}{2\sigma}} = \frac{r N}{2a}\sqrt{\frac{\mu_o \omega}{2\sigma}} \tag{6.10}$$

where the total length l of the wire with radius of a can be estimated as $2\pi r * N$. Other than ohmic loss, there is power loss in the radiation resistance, which is given by

$$R_{radiative} = \sqrt{\frac{\mu_o}{\epsilon_o}}\left[\frac{\pi N^2}{12}\left(\frac{\omega r}{c}\right)^4 + \frac{2}{3\pi^3}\left(\frac{\omega h}{c}\right)^2\right] \tag{6.11}$$

According to Kurs et al. [163], the first term in Equation 6.11 is a magnetic dipole radiation term (assuming $r \ll 2\pi c/\omega$, where c is the speed of light), and the second term is due to the electric dipole of the coil. The second term is much smaller than the first term for these WPT system parameters, so the second term of Equation 6.11 is ignored for simplicity. Hence, by substituting for ω and c, the radiative resistance is simplified to

$$R_{radiative} = \sqrt{\frac{\mu_o}{\epsilon}}\left[\frac{4\pi^5 N^2}{3}\left(\frac{r}{\lambda}\right)^4\right] = 15600\ \pi^2 N^2 \left(\frac{r}{\lambda}\right)^4 \tag{6.12}$$

As mentioned, for a system to operate in the strongly coupled regime, the term κ^2/Γ^2 must be greater than 1 to achieve an efficient WPT system. Referring to Equations 6.7 and 6.9, the coupling-to-loss ratio κ/Γ can be expressed as follows:

$$\frac{\kappa}{\Gamma} = \frac{\omega M L}{\left(R_{ohmic} + R_{radiative}\right)\sqrt{L_1 L_2}} \tag{6.13}$$

The basic configuration of the transmitting and receiving coils is assumed to be the same dimensions, and the WPT system is in tuned resonance. By doing so, Equation 6.13 can be further elaborated, by substituting $L_1 = L_2$ and the radius of both coils to be equal, $r = r_b$, as follows:

$$\frac{\kappa}{\Gamma} = \frac{\mu_0 \pi \omega r^3}{(r^2 + D^2)^{1.5}} \left[\frac{N}{\frac{1}{a}\sqrt{\frac{\mu_0 \omega}{2\sigma}} + \frac{\pi N r^3 \omega^4}{6c^4}\sqrt{\frac{\mu_0}{\epsilon_0}}} \right] \tag{6.14}$$

Referring to Equation 6.14, it can be observed that there are some prominent factors in the equation that have effects on the coupling-to-loss ratio, hence the efficiency of the WPT system. These factors include the conductivity and radius of the wire, the coil size (i.e., radius and number of turns), the distance between two coils, and the operating frequency of the WPT system. As can be seen in Equation 6.14, the equation is highly interdependent on various design factors; hence, there is a need to find an optimum configuration for the WPT system. By conducting simulations and experiments, the relationships between these design factors and the efficiency of the system are determined.

6.2.2 Simulation Results

In order to improve the efficiency of the WPT system, four different simulations were conducted to determine the relationships between the efficiency of the system and the design factors: (1) frequency f, (2) coil radius r, (3) number of winding turns in a coil N, and (4) distance D between two coils [173]. All the simulations were based on the efficiency and coupling-to-loss ratio expressed in Equations 6.6 and 6.14, respectively.

6.2.2.1 Simulation of Efficiency versus Frequency

The basic configuration of the two cylindrical coils used in this simulation was defined as follows: The coil radius r of 7.5 cm was made of 5 winding turns N of copper wire (SWG12) with conductor radius a of 0.13208 cm and conductivity of $59.6*10^6$ S/m. The separation distance D between the two coils was 10 cm. The efficiency η of the WPT system was simulated over a range of operating frequencies from 1 Hz to 1000 GHz, and the simulation results are plotted in Figure 6.16.

It is observed from Figure 6.16 that there exists a band of frequencies between 1 and 1000 GHz whereby the efficiency of the WPT system was optimal. This band of optimum efficiency is the strongly coupled regime where $\kappa/\Gamma \gg 1$ and maximum power transfer took place. As the operating frequency shifted away from the strongly coupled regime, the WPT system efficiency started to decrease. At the lower-frequency region, the coupling between the coils became weak, as illustrated in Equation 6.7. Similarly, at the higher-frequency region, the power losses in the ohmic and radiative resistances became prominent, as illustrated in Equation 6.9. Only when the coupling-to-loss ratio balances will the efficiency of the WPT system be high.

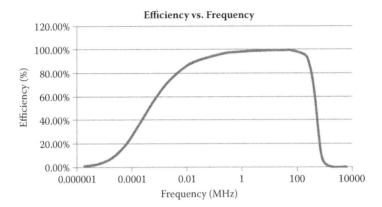

FIGURE 6.16
Efficiency of a WPT system for various operating frequencies.

6.2.2.2 Simulation of Efficiency versus Coil Radius

Like the previous simulation, the same basic configuration of the two coils was used except that the radius of the coils was changed between 5 and 100 cm. The range of different radii was chosen in conjunction with the needs of various commercial applications, like miniaturized biomedical devices to electric vehicles. For each coil radius, the efficiency curve is swept across a frequency range of 1 to 1000 MHz as shown in Figure 6.17.

Referring to Figure 6.17, as the radius of the coil increased, it is observed that the optimum efficiency band shifted towards the left of the figure and widened. From this observation, it is notable that for any frequency between

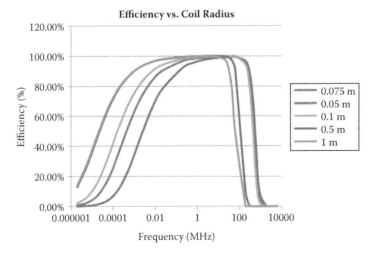

FIGURE 6.17
Efficiency of a WPT system for various coil radii.

1 and 50 MHz, the efficiency of the WPT system still remained at its optimal for different coil radii between 5 and 100 cm. Take, for example, at operating frequency of 100 kHz, it is observed from Figure 6.17 that the efficiencies of the WPT system with a coil radii of 5 and 100 cm were 60% and 95% respectively. There was a significant difference in the system efficiency of around 35% when the coil radius changed. As such, it is required to operate the WPT system with a coil radius of 5 to 100 cm to be within 1 to 50 MHz so that maximum power throughput is achieved.

6.2.2.3 Simulation of Efficiency versus Number of Turns

In this case, the number of turns in each coil were varied for 3, 5, and 7 turns to determine the behaviour of the WPT system efficiency over a range of frequencies. According to Equation 6.8, it can be seen that the mutual inductance M of the coils is a function of the square of N turns. As N increases, it is expected that M increases and so does the coupling coefficient κ, where the relationship is illustrated in Equation 6.7. The simulated results are shown in Figure 6.18.

Like the previous case, it is observed in Figure 6.18 that when the number of turns changed from 3 to 7, the optimum efficiency band starts to shift again towards the lower-frequency region. This is because of the change in the ohmic and radiative resistances, as expressed in Equations 6.10 and 6.11, respectively, as well as the coupling coefficient κ, which ended up changing the efficiency of the WPT system.

6.2.2.4 Simulation of Efficiency versus Distance

The mutual inductance of the coils, as expressed in Equation 6.8, is inversely proportional to the cubic separation distance D between the coils. As the

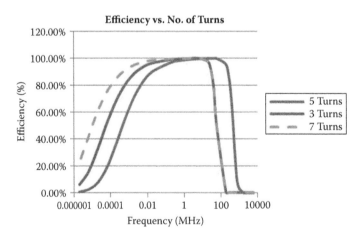

FIGURE 6.18
Efficiency of a WPT system for different numbers of winding turns.

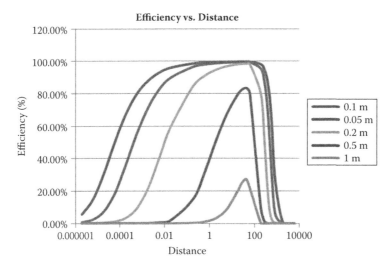

FIGURE 6.19
Efficiency of a WPT system for different separation distances.

distance between the two coils increases, the mutual coupling between the coils decreases as illustrated by Equation 6.7. The distance factor has no effect on the other performance indicators of the WPT system. In this simulation, the distance between the two coils varied from 5 to 100 cm, and the efficiency of the system over a range of frequencies of 1 to 1000 GHz is plotted in Figure 6.19.

With reference to Figure 6.19, it is observed that for shorter distances (i.e., 5 and 10 cm), the optimum efficiency band remained more or less the same due to the strong coupling between the coils. As the separation distance started to increase, the efficiency of the WPT system decreased. At a distance of 50 cm, the peak attainable efficiency of the system dropped to around 80%. This reduction in the efficiency of the WPT system continued and it dropped to around 30% when the separation distance between the coils was lengthened to 100 cm, which was twice of its counterpart of 50 cm. After conducting the simulations, the relationship between the WPT efficiency and the design factors was known, and it is thus possible to leverage the understanding onto the experimental design of an optimal WPT system.

6.2.3 Characteristics of the WPT System

Experiments were conducted to observe the performance of the WPT system with respect to the three more prominent design parameters: (1) frequency, (2) distance, and (3) load. The rest of the design parameters were also investigated; however, their effects were not as significant as the three chosen parameters.

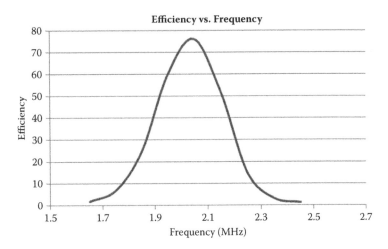

FIGURE 6.20
Experimental efficiency of a WPT for various operating frequencies.

6.2.3.1 Experimental Efficiency versus Frequency

Referring to the simulation shown in Section 6.2.2, the simulation results showed that the efficiency of the WPT system was maximum for an optimal operating frequency. This optimum efficiency was the strongly coupled regime where $\kappa/\Gamma \gg 1$ and maximum power transfer took place. As the operating frequency shifted away from the strongly coupled regime, the WPT system efficiency started to decrease. The same deduction is also achieved, and its experimental results are plotted in Figure 6.20.

Figure 6.20 verifies that the resonant frequency of the WPT system of 2.05 MHz was well within the simulated frequency band of 1 to 40 MHz. At resonance, the maximum obtainable efficiency of the WPT system was around 75%. Shifting away from the resonant frequency of 2.05 MHz by around ±10%, it can be observed from Figure 6.20 that the WPT efficiency dropped tremendously to less than 15%. It is necessary to design the WPT coils to operate at resonance using Equation 6.2 to achieve efficient wireless power transmission.

6.2.3.2 Experimental Efficiency versus Distance

In this experiment, the operating frequency of the power source was set at the resonance of the WPT system of 2.05 MHz, and the separation distance between the transmitting and receiving coils was tested over a transmission range of 40 cm. The experimental results are collected and plotted in Figure 6.21.

Referring to Figure 6.21, it was observed that the efficiency of the WPT system was maximum at a separation distance of 20 cm. Beyond the distance of 20 cm where the coils are separated even further, the mutual coupling between the transmitting and receiving coils became weaker and weaker, so less and

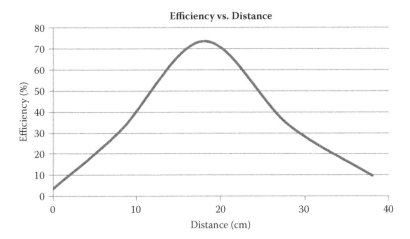

FIGURE 6.21
Experimental efficiency of a WPT over a range of separation distances.

less electrical power was transferred wirelessly over to the load. Conversely, when the two coils became closer to each other, the mutual coupling between the coils became stronger; at the same time, the counter-emf effect rose. The counter-emf is the voltage, or electromotive force, that pushes against the current that induces it. It is caused by a changing electromagnetic field, which is represented by Lenz's law of electromagnetism. The voltage's polarity is at every moment the reverse of the input voltage. When a rapidly changing magnetic field induces an emf in a coil, a current caused by this emf flows. This current flow would in turn generate a magnetic field in the coil that opposes the original magnetic field that created it, and this would ultimately reduce the induced emf in the coil. As such, it can be seen in Figure 6.21 that as the distance between the two coils decreased, the efficiency of the WPT system also decreased due to the counter-emf effect.

6.2.3.3 Experimental Efficiency versus Load

Once the resonant frequency and separation distance of the WPT system were fixed at 2.05 MHz and 20 cm, respectively, a load resistance ranging from 10 Ω to 10 kΩ was connected to the load coil to determine the characteristic of the WPT system. Figure 6.22 shows the efficiency plot of the WPT system under different loading conditions.

Referring to Figure 6.22, it is observed that a maximum WPT efficiency of 75% was attainable at a matching load resistance of 220 Ω. However, for other loading conditions, shifting away from the internal resistance of the load coil of the WPT system, either very light or heavy electrical loads, the efficiency of the system dropped significantly.

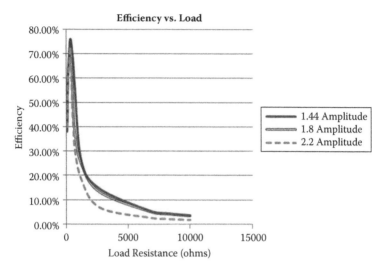

FIGURE 6.22
Experimental efficiency of a WPT under different loading conditions.

6.2.4 Experimental Results

The experimental setup of the WPT system, as shown in Figure 6.23, consisted of a high-power, high-frequency AC source; a set of source, transmitting, receiving, and load coils; electrical testing loads; and an oscilloscope. To achieve both high-power and high-frequency electrical supply from the AC source is very challenging, but it has been successfully implemented using a low-power, high-frequency signal generator capable of generating an AC signal up to 500 MHz, and a high-power amplifier. The high-frequency AC signal generated by the signal generator was channelled to the power amplifier for amplification.

6.2.4.1 WPT System Powering Electrical Load(s)

The designed WPT system was experimentally tested with an electrical appliance as the system load instead of a resistor to demonstrate and determine the wireless power capability of the system. To add a physical perspective to the WPT research work, a 12-V lightbulb was used. Figure 6.24 shows a 12-V lightbulb lit at a distance of 20 cm between the transmitting and receiving coils.

Based on the experimental results, the WPT system shown in Figure 6.24 was able to transmit an electrical output power of 1 W over a distance of 20 cm to the lightbulb with an efficiency of around 51%. During the experiment, whenever the coil separation distance was beyond the limit of 20 cm, the brightness of the lightbulb diminished quickly. This phenomenon was due to the low efficiency of the WPT system with weak coupling between the coils. The same experiment was conducted for different lightbulbs of 2.4, 3.6, and 7.2 V as well, and their experimental results, which include input power

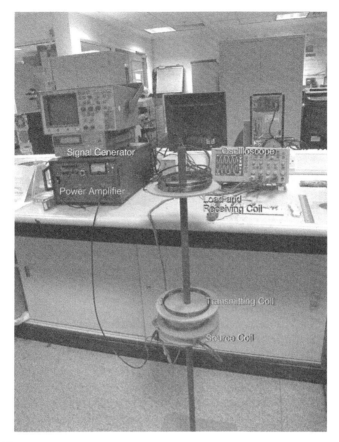

FIGURE 6.23
Experimental setup of the WPT system.

P_{in} at the source coil, output power P_{out} at the load coil, and efficiency η_{WPT} of the WPT system, are recorded in Table 6.2.

Referring to Table 6.2, it can be observed that the output power and efficiency of the WPT system for different loading conditions (i.e., 2.4 to 12-V lightbulbs) ranged between 0.6 and 1 W and 35% to 50%, respectively. One interesting observation to note from Table 6.2 is that for the 2.4-V lightbulb, as the distance between the coils was reduced from 20 to 15 cm, the WPT efficiency increased by more than a third to around 47.5%. This positive observation is in line with the conclusion from the discussion of simulation in Section 6.2.2. Once the one-to-one WPT system was investigated, the following experiment shown in Figure 6.25 was designed to demonstrate the concept of powering multiple devices, that is, LED load 1 and lightbulb load 2 of different geometries operating at the same resonant frequency.

With reference to Figure 6.25, it can be seen that a single source and transmitting coil were able to power up two separate load and receiving coils 1

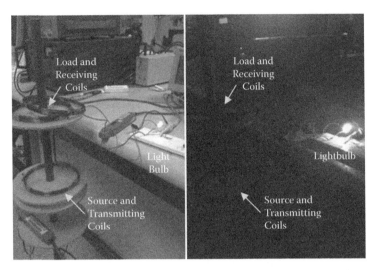

FIGURE 6.24
Demonstration of wireless power transmission of watt level.

and 2 of different dimensions tuned to the same frequency by adjusting their capacitances based on the fundamental equation given by Equation 6.2. Both the LED and lightbulb loads were lit with the same WPT source, and this exhibited the capability of the WPT system to transfer power to multiple electrical devices in a wireless manner. Another interesting observation to note from Figure 6.25 is that the smaller coil (i.e., load and receiving coil 1), placed on the table was out of the source coil's line of sight, but magnetic waves were still able to power the load without interacting with the extraneous objects, such as the wood and steel of the table, between the transmitting and receiving coils. This is one of the unique advantages of using resonant magnetic waves as they do not interact with nonmagnetic materials such as plastic and wood. Even if there was magnetic material between the transmitting and receiving coils, as can be seen in Figure 6.25, where load and receiving coil 1 is placed on the table made of wood and steel, energy was still able to be transferred wirelessly to the LED load 1. This is because the fundamental

TABLE 6.2

Efficiency of WPT System Powering Different Lightbulbs

Bulb Rating (V)	Distance (cm)	V_{in} (V)	I_{in} (A)	V_{out} (V)	I_{out} (A)	P_{in} (W)	P_{out} (W)	η_{WPT} (%)
2.4	15	5.9	0.12	1.05	0.32	0.708	0.336	47.5
2.4	20	6.5	0.28	1.64	0.39	1.820	0.640	35.2
3.6	20	7.2	0.29	2.05	0.39	2.088	0.800	38.3
7.2	20	8.1	0.26	2.40	0.35	2.106	0.840	39.9
12	20	8.5	0.25	3.17	0.34	2.125	1.078	50.7

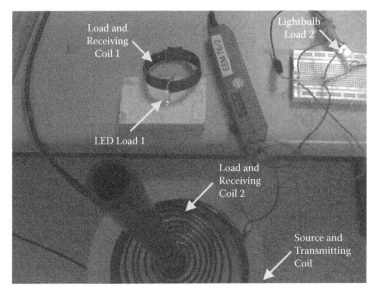

FIGURE 6.25
Powering multiple loads and effect of extraneous objects.

nature of resonance dictates that resonant objects interact very weakly with other nonresonant extraneous objects such that very little electrical energy is lost.

6.2.4.2 Network of WPT Resonator Coils

Another objective of this WPT research was to network the designed high-efficiency WPT resonator coils together to achieve the ultimate goals of wireless and batteryless electronic products. The designed resonator coils were networked together, as illustrated in Figure 6.26, such that each resonator coil in the network received and relayed electrical power from and to its neighbouring resonator coil in accordance with the WPT routes illustrated by dotted lines in Figure 6.26.

Referring to Figure 6.26, the mobile gadgets (e.g., camera, iPhone, laptop, etc.) were equipped with resonator coils, and whenever these mobile gadgets entered the room or rather the WPT network, they would be charged. In the midst of charging, the resonator coil also played the role of relaying the received electrical power to its nearby mobile gadget. By doing so, it was possible to reduce the wireless power transmission distance between each gadget; hence, the form factor of the resonator coil could be smaller to fit into the mobile gadget. It was thus more viable to utilize WPT technology as a self-sustaining portable energy source for the mobile gadgets. Figure 6.27 shows the powering of an LED at a separation distance of 60 cm away from the source and transmitting coil by using a network of resonator coils (i.e., resonators #1 and #2).

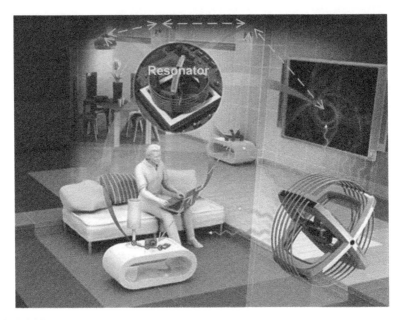

FIGURE 6.26
Resonators in network form to receive and relay electrical power.

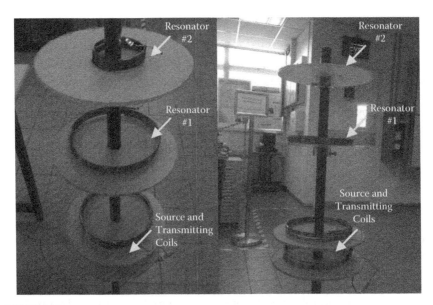

FIGURE 6.27
A resonant coil receives and relays electrical power to another nearby resonator coil.

As compared to the experiment conducted in Section 6.2.4 where one pair of transmitting and receiving coils was used, it was observed that if an additional resonating coil (i.e., resonator #1) was placed between the transmitting and receiving coils, the transmission range of the WPT system increased from 20 to 60 cm. Referring to Figure 6.27, it can be observed that the intermediate resonator coil #1 acted as a source for the next resonator coil #2 as well as a relay energy source for the subsequent resonator coils. Although more resonator coils were used to extend the wireless power transmission distance, the optimal design of the WPT system should take note of the additional ohmic loss in each hop. Hence, the experiment successfully demonstrated that a network of resonator coils can be used to transfer electrical power efficiently over a larger wireless transmission range to meet the target application.

6.2.5 Summary

The concept of using self-resonant coils operating in the strongly coupled regime was experimentally demonstrated. It is obvious that this process was far superior to simple inductive coupling. The equations and theory behind the concept were analyzed, and further derivations and simulations helped in designing the appropriate system. Various design factors, such as resonant frequency, load, and distance, were changed and subjected to experiments. The results obtained were consistent with the derived equations and simulations. Experimental results showed that an electrical output power of 1 W was successfully delivered to the load, which was a lightbulb at an efficiency of 51% at a 20-cm coil separation distance with coils of 7.5-cm radius.

7

Conclusions and Future Works

This chapter concludes the book; it briefly restates the motivation of the research works recorded in the book, the identified problem areas, and the various findings in each problem area. Finally, it shows the direction of future research in this regard.

7.1 Conclusions

This book covers work done on design, analysis, and practical implementation of various optimized energy harvesting (EH) systems for sustaining self-autonomous wireless sensor nodes. Conventionally, these tiny, smart, and inexpensive sensor nodes, connected together into a wireless sensor network (WSN), are powered by alkaline or rechargeable batteries. They are scattered in the targeted deployment field to facilitate monitoring and controlling of physical environments from remote locations that are too difficult or dangerous to reach. However, as the coverage area of the WSN becomes larger, dense with many sensor nodes, problems arise in powering each of these battery-operated sensor nodes. The onboard batteries of the sensor nodes have limited energy capacity, so after operating the sensor nodes for some time the batteries deplete and the sensor nodes go into an idle state. Hence, the limitation of energy sources for sensor nodes becomes critical, even worse when one considers the prohibitive cost of providing power through wired cables to them or replacing batteries. Furthermore, when the sensor nodes must be extremely small, as tiny as several cubic centimetres, to be conveniently placed and used, such small volumetric devices are very limited in the amount of energy that the batteries can store, and there would be severe limits imposed on the nodes' lifetime powered by the miniaturized battery that is meant to last the entire life of the node.

Energy harvesting emerges as a highly potential solution to make a paradigm shift from battery-operated conventional, the WSN, towards a truly self-autonomous and sustainable energy harvesting wireless sensor network (EH-WSN). The main focus of this research is to design and implement EH systems to resolve the energy supply problems faced by the wireless sensor nodes. In an EH system, there are generally four main parts; energy harvester (source), power management circuit, energy storage device, and wireless sensor node (load). Power output per unit mass or volume (i.e., power/energy density) is a key performance unit for the energy harvester. Based on the characteristic and capability of each of the ambient energy sources, the EH system is designed to suit the target application and its ambient conditions and event/task requirements. For the EH-WSN, various types of ambient renewable EH systems, based on wind energy harvesting (WEH), thermal energy harvesting, vibration energy harvesting (VEH), solar energy harvesting, hybrid energy harvesting (HEH), and magnetic energy harvesting are designed and then implemented into hardware prototypes for proof of concept. To optimize these EH systems, the harvested power is conditioned to an appropriate form for either charging the system batteries or powering the connected load directly. Proper load impedance matching between the EH source and the electrical load is carried out to maximize the usage of the harvested energy. Several different types of power-electronic-based management circuits, such as an active alternating current-direct current (AC-DC) converter, DC-DC converter with maximum power point tracking, energy storage and latching circuit, and others, have been introduced and implemented.

7.2 Future Research Works

In this EH research work, starting from scrap to what has been achieved so far, a wide breadth of studies and depth investigations on critical issues have been carried out. However, there is still room for more research, improvements, and further optimization. Some possible future research works include those discussed next.

For the direct WEH research, the efficiency of the electric generator is relatively low because the generator is not optimized to operate at its rated speed. To overcome that, several ways have been suggested to improve the efficiency of the electric generator, like introducing a gearbox between the wind turbine generator (WTG) hub and the generator and modifying the two-pole generator to a multipole generator. Other than the electric generator, the wind turbine itself is another possible area for future research. With adequate design of the blades of the wind turbine to suit the wind profile at the point of deployment, the aerodynamic efficiency can be improved. As for indirect WEH research, the analysis for the conversion process from wind power to electrical power through the piezoelectric cantilever beam effect can be enhanced using computer-aided engineering software such as Ansys. By doing so, with prior knowledge of the incoming wind speed, the power throughput of the vibration-based piezoelectric wind energy harvester can be estimated. In addition, the mechanical resonance of the harvester's structure can be optimized to suit the needs of a target application.

In the VEH system, the energy extracted by the impact-based piezoelectric energy harvester is sufficient to power the radio-frequency (RF) transmitter load. However, only a small fraction of the harvested energy has been extracted by the vibration energy harvester for use. As such, for future research work, the energy extraction process can be optimized using dedicated power management techniques like synchronized electric charge extraction (SECE), parallel or series synchronized switch harvesting on inductor (SSHI), and so on. In addition, further investigations are necessary for commercially viable, cheaper piezoelectric material with similar high performance. This step is necessary for the commercialization of this impact-based VEH research work.

Hybrid energy harvesting from two renewable energy sources has been proven with a hardware prototype to yield more electrical power than a single EH source. The augmenting process of two energy sources is done by either using an individual power converter for each EH source or directly connecting the energy sources in parallel. Other than the parallel configuration, for future research, the renewable energy sources can be stacked in series to produce higher voltage but at the same time yield more electrical power. In addition, it is also worthwhile to explore EH from more than two energy sources available in the ambient environment. With more electrical energy harvested from different energy sources, the reliability of the wireless sensor nodes can be improved.

Last, in this book, the concept of magnetic energy harvesting via inductive coupling has been presented and demonstrated. An adequate power management circuit was developed to bridge between the magnetic energy harvester and the electrical load. However, there was no proper matching between the impedances of the source and the load. As such, in future research work, an intelligent control of the power electronic converter to match the source and load impedances can be developed. With this proper impedance-matching scheme in place, the harvested power from the magnetic energy harvesting would be maximum. This surplus in harvested energy would be useful for the operation of the electrical load. For the research work on wireless power transfer (WPT) via strongly coupled magnetic resonances, it is clear that evanescent resonant coupling is able to transmit a watt level of electrical power from the transmitting end to the receiving end over a distance of 1 to 2 m. In order to further increase the power levels beyond a few watts, there are two factors to be considered: (1) The power rating of the matching capacitors has to be increased. Capacitors using suitable material like tantalum that can withstand high current and operate at the megahertz frequency range can be used to replace ceramic capacitors. (2) High-frequency, high-power amplifiers to generate an AC electrical power in the megahertz range and watt level of power can be used.

References

1. A. Kansal, J. Hsu, S. Zahedi, and M.B. Srivastava, "Power management in energy harvesting sensor networks," *ACM Transactions on Embedded Computing Systems*, p. 35, 2006.
2. D.J. Cook and S.K. Das, "Wireless sensor networks," *Smart Environments: Technologies, Protocols and Applications*, Wiley, New York, 2004.
3. Technology Review, "10 emerging technologies that will change the world," February 2003 issue of *Technology Review*, http://www.technologyreview.com/Infotech/13060/?a=f, accessed on June 7, 2010.
4. I.F. Akyildiz, W.L. Su, S. Yogesh, and C. Erdal, "A survey on sensor networks," *IEEE Communications Magazine*, vol. 40, no. 8, pp. 102–114, 2002.
5. K. Sohrabi, J. Gao, V. Ailawadhi, and G. Pottie, "Protocols for self-organization of a wireless sensor network," *IEEE Personal Communications*, vol. 7, no. 5, pp. 16–27, 2000.
6. Tsung-Hsien Lin, W.J. Kaiser, and G.J. Pottie, "Integrated low-power communication system design for wireless sensor networks," *IEEE Communications Magazine*, vol. 42, no. 12, pp. 142–150, 2004.
7. A. Sinha and A. Chandrakasan, "Dynamic power management in wireless sensor networks," *IEEE Design and Test of Computers*, March–April, pp. 62–74, 2001.
8. G.V. Merrett, B.M. Al-Hashimi, N.M. White, and N.R. Harris, "Resource aware sensor nodes in wireless sensor networks," *Journal of Physics*, vol. 15, no. 1, pp. 137–142, 2005.
9. E. Lattanzi, E. Regini, A. Acquaviva, and A. Bogliolo, "Energetic sustainability of routing algorithms for energy-harvesting wireless sensor networks," *Computer Communications*, vol. 30, nos. 14–15, pp. 2976–2986, 2007.
10. C. Chong and S.P. Kumar, "Sensor networks: Evolution, opportunities, and challenges," *Proceeding of the IEEE, Sensor Networks and Applications*, vol. 91, no. 8, pp. 1247–1256, 2003.
11. M. Kuorilehto, M. Hannikainen, and T.D. Hamalainen, "A survey of application in wireless sensor networks," *EURASIP Journal on Wireless Communications and Networking*, vol. 2005, no. 5, pp. 774–788, 2005.
12. M. Tubaishat and S. Madria, "Sensor networks: an overview," *Potentials, IEEE*, vol. 22, pp. 20–23, 2003.
13. Edgar H. Callaway, *Wireless Sensor Networks: Architectures and Protocols*, Auerbach, Boca Raton, FL, 2003.
14. D. Culler, D. Estrin, and M. Srivastava, "Overview of sensor networks," *IEEE Computer*, vol. 37, no. 8, pp. 41-49, 2004.
15. N. Kurata, S. Saruwatari, H. Morikawa, "Ubiquitous structural monitoring using wireless sensor networks," *International Symposium on Intelligent Signal Processing and Communications*, pp. 99–102, 2006.
16. J.L. Hill and D.E. Culler, "Mica: a wireless platform for deeply embedded networks," *IEEE Micro*, vol. 22, pp. 12–24, 2002.

17. J.M. Rabaey, M.J. Ammer, J.L. da Silva, Jr., D. Patel, and S. Roundy, "PicoRadio supports ad hoc ultra-low power wireless networking," *IEEE Computer*, vol. 33, pp. 42–48, 2000.

18. Massachusetts Institute of Technology (MIT), "μAmps projects," Microsystems Technology Laboratories, http://www-mtl.mit.edu/researchgroups/icsystems/uamps/, accessed on June 7, 2010.

19. TinyOS, "The TinyOS Project," TinyOS Community Forum, http://www.tinyos.net, accessed on June 7, 2010.

20. R. Verdone, D. Dardari, G. Mazzini, and A. Conti, "network lifetime," *Wireless Sensor and Actuator Networks: Technologies, Analysis and Design*, Academic Press, New York, pp. 115–116, 2008.

21. Crossbow Technology Inc. (California), "MPR-MIB Users Manual," Crossbow Resources, Revision A, 2007.

22. J.W. Tester, "Energy transfer and conversion methods," *Sustainable Energy Lecture Notes, Topic on Energy Storage Modes*, MIT, Cambridge, MA, 2005.

23. S.F.J. Flipsen et al., "Alternative power sources for portables and wearable. Part 1—power generation and Part 2—Energy storage," Technical Report, Personal Energy Systems Programme, Delft University of Technology, Delft, Netherlands, 2004.

24. G.E. Blomgren, "Perspectives on portable lithium ion batteries liquid and polymer electrolyte types," *17th Annual Battery Conference on Applications and Advances*, pp. 141–144, 2002.

25. V. Raghunathan, S. Ganeriwal, and M. Srivastava, "Emerging techniques for long-lived wireless sensor networks," *IEEE Communications Magazine*, vol. 44, no. 4, pp. 108–114, 2006.

26. D. Niyato, E. Hossain, M.M. Rashid, and V.K. Bhargava, "Wireless sensor networks with energy harvesting technologies: A game-theoretic approach to optimal energy management," *IEEE Wireless Communications*, vol. 14, no. 4, pp. 90–96, 2007.

27. J.P. Thomas, M.A. Qidwai, and J.C. Kellogg, "Energy scavenging for small-scale unmanned systems," *Journal of Power Sources*, vol. 159, pp. 1494–1509, 2006.

28. J.A. Paradiso and T. Starner, "Energy scavenging for mobile and wireless electronics," *IEEE Pervasive Computing*, vol. 4, no. 1, pp. 18–27, 2005.

29. C. Mathna, T. O'Donnell, R.V. Martinez-Catala, J. Rohan, and B. O'Flynn, "Energy scavenging for long-term deployable wireless sensor networks," *Talanta*, vol. 75, no. 3, pp. 613–623, 2008.

30. J.F. Randall and J. Jacot, "Is AM1.5 applicable in practice? Modelling eight photovoltaic materials with respect to light intensity and two spectra," *Renewable Energy*, vol. 28, no. 12, pp. 1851–1864, 2003.

31. M.A. Green, K. Emery, Y. Hisikawa, and W. Warta, "Short communication solar cell efficiency tables (version 30)," *Progress in Photovoltaics: Research and Applications*, vol. 15, no. 5, pp. 425–430, 2007.

32. V. Raghunathan, A. Kansal, J. Hsu, J. Friedman, and M. Srivastava, "Design considerations for solar energy harvesting wireless embedded systems," *4th International Symposium on Information Processing in Sensor Networks (IPSN)*, pp. 457–462, 2005.

33. X.F. Jiang, J. Polastre, and D.E. Culler, "Perpetual environmentally powered sensor networks," *4th International Symposium on Information Processing in Sensor Networks (IPSN)*, pp. 463–468, 2005.

34. F.I. Simjee and P.H. Chou, "Efficient charging of supercapacitors for extended lifetime of wireless sensor nodes," *IEEE Transactions on Power Electronics*, vol. 23, no. 3, pp. 1526–1536, 2008.

35. C. Park and P.H. Chou, "AmbiMax: Autonomous energy harvesting platform for multi-supply wireless sensor nodes," *3rd Annual IEEE Communications Society on Sensor and Ad Hoc Communications and Networks (SECON)*, vol. 1, pp. 168–177, 2006.

36. D. Dondi, A. Bertacchini, D. Brunelli, L. Larcher, and L. Benini, "Modeling and optimization of a solar energy harvester system for self-powered wireless sensor networks," *IEEE Transaction on Industrial Electronics*, vol. 55, no. 7, pp. 2759–2766, 2008.

37. N.S. Hudak and G.G. Amatucci, "Small-scale energy harvesting through thermoelectric, vibration, and radio frequency power conversion," *Journal of Applied Physics*, vol. 103, no. 10, pp. 101–301(1-24), 2008.

38. F. Cottone, "Nonlinear piezoelectric generators for vibration energy harvesting," Ph.D. thesis, University of Perugia, Italy, 2008.

39. V. Leonov, T. Torfs, P. Fiorini, and C. Van Hoof, "Thermoelectric converters of human warmth for self-powered wireless sensor nodes," *IEEE Sensors Journal*, vol. 7, no. 5, pp. 650–657, 2007.

40. H.A. Sodano, G.E. Simmers, R. Dereux, and D.J. Inman, "Recharging batteries using energy harvested from thermal gradients," *Journal of Intelligent Material Systems and Structures*, vol. 18, no. 1, pp. 3–10, 2007.

41. T. Kanesaka, "Development of a thermal energy watch," *Proceedings of the 64th Conference on Chronometry (Socit Suisse de Chronomtrie)*, Le Sentier, Switzerland, pp. 19–22, 1999.

42. J.W. Stevens, "Heat transfer and thermoelectric design considerations for a ground-source thermo generator," *Proceedings of 18th International Conference on Thermoelectrics*, 1999.

43. E.E. Lawrence and G.J. Snyder, "A study of heat sink performance in air and soil for use in a thermoelectric energy harvesting device," *Proceedings of 21st International Conference on Thermoelectrics (ICT 02)*, 2002.

44. C. Moser, "Power management in energy harvesting embedded systems," Ph.D. thesis, Swiss Federal Institute of Technology, Zurich, 2009.

45. S. Roundy, P.K. Wright, and J.M. Rabaey, *Energy Scavenging for Wireless Sensor Networks with Special Focus on Vibrations*, Kluwer Academic Press, Boston, 2004.

46. N. Shenck and J. Paradiso, "Energy scavenging with shoe-mounted piezoelectrics," *IEEE Micro*, vol. 21, no. 3, pp. 30–42, 2001.

47. S.J. Roundy, "Energy scavenging for wireless sensor nodes with a focus on vibration to electricity conversion," Ph.D. thesis, University of California, Berkeley, 2003.

48. P. Glynne-Jones, M.J. Tudor, S.P. Beeby, and N.M. White, "An electromagnetic, vibration-powered generator for intelligent sensor systems," *Sensors and Actuators*, vol. 110, nos. 1–3, pp. 344–349, 2004.

49. J. Edmison, M. Jones, Z. Nakad, and T. Martin, "Using piezoelectric materials for wearable electronic textiles," *Proceedings of 6th International Symposium on Wearable Computers (ISWC)*, 2002.

50. S. Meninger, A.P. Amirtharajan, and R. Chandrakasan, "Vibration-to-electric energy conversion," *IEEE Transaction on VLSI System*, vol. 9, pp. 64–71, 2001.

51. P.D. Mitcheson, T.C. Green, E.M. Yeatman, and A.S. Holmes, "Architectures for vibration-driven micropower generators," *Journal of Microelectromechanical Systems*, vol. 13, no. 3, pp. 429–440, 2004.

52. J.A. Paradiso and Mark Feldmeier, *A Compact, Wireless, Self-Powered Pushbutton Controller*, MIT Media Laboratory, Cambridge, MA, 2002.

53. E. Braunwald, *Heart Disease: A Textbook of Cardiovascular Medicine*, Saunders, Philadelphia, 1980.

54. M. Ramsay and W. Clark, "Piezoelectric energy harvesting for bio-MEMs applications," *Proceedings of the SPIE—The International Society for Optical Engineering*, vol. 4332, pp. 429–439, 2001.

55. Z. Chen, J.M. Guerrero, and F. Blaabjerg, "A review of the state of the art of power electronics for wind turbines," *IEEE Transactions on Power Electronics*, vol. 24, no. 8, pp. 1859–1875, 2009.

56. S. Heier (author) and R. Waddington (translator), *Grid Integration of Wind Energy Conversion Systems*, 2nd ed., Wiley, Chichester, England, 2006.

57. Renewable Resource Data Center (RReDC), National Renewable Energy Laboratory, http://www.nrel.gov/rredc/, accessed on June 14, 2010.

58. C.T. Chen, R.A. Islam, and S. Priya, "Electric energy generator," *IEEE Transactions on Ultrasonics, Ferroelectrics and Frequency Control*, vol. 53, no. 3, pp. 656–661, 2006.

59. A.S. Holmes, G. Hong, K.R. Pullen, and K.R. Buffard, "Axial-flow microturbine with electromagnetic generator: Design, CFD simulation, and prototype demonstration," *17th IEEE International Conference on Micro Electro Mechanical Systems*, pp. 568–571, 2004.

60. M.A. Weimer, T.S. Paing, and R.A. Zane, "Remote area wind energy harvesting for low-power autonomous sensors," *37th IEEE Power Electronics Specialists Conference*, pp. 2911–2915, 2006.

61. S. Priya, C.T. Chen, D. Fye, and J. Zahnd, "Piezoelectric windmill: A novel solution to remote sensing," *Japanese Journal of Applied Physics*, vol. 44, no. 3, pp. L104–L107, 2005.

62. R. Myers, M. Vickers, K. Hyeoungwoo, and S. Priya, "Small-scale windmill," *Applied Physics Letters*, vol. 90, no. 5, p. 54106-1-3, 2007.

63. D. Maurath, C. Peters, T. Hehn, M. Ortmanns, and Y. Manoli, "Highly efficient integrated rectifier and voltage boosting circuits for energy harvesting applications," *Advanced Radio Science*, vol. 6, pp. 219–225, 2008.

64. Y.H. Lam, W.H. Ki, and C.Y. Tsui, "Integrated low-loss CMOS active rectifier for wirelessly powered devices," *IEEE Transactions on Circuits and Systems II: Express Briefs*, vol. 53, no. 12, pp. 1378–1382, 2006.

65. M.D. Seeman, S.R. Sanders, and J.M. Rabaey, "An ultra-low-power power management IC for energy-scavenged wireless sensor nodes," *IEEE Power Electronics Specialists Conference (PESC 2008)*, pp. 925–931, 2008.

66. E. Koutroulis and K. Kalaitzakis, "Design of a maximum power tracking system for wind-energy-conversion applications," *IEEE Transactions on Industrial Electronics*, vol. 53, no. 2, pp. 486–494, 2006.

67. Z. Chen, and E. Spooner, "Grid interface options for variable-speed, permanent-magnet generators," *IEEE Proceedings—Electronic Power Applications*, vol. 145, no. 4, pp. 273–283, 1998.

68. Q. Wang and L. Chang, "An intelligent maximum power extraction algorithm for inverter-based variable speed wind turbine systems," *IEEE Transactions on Power Electronics*, vol. 19, no. 5, pp. 1242–1249, 2004.

69. K. Khouzam and L. Khouzam, "Optimum matching of direct-coupled electrome-chanical loads to a photovoltaic generator," *IEEE Transactions on Energy Conversion*, vol. 8, no. 3, pp. 343–349, 1993.

70. T. Paing, J. Shin, R. Zane and Z. Popovic, "Resistor emulation approach to low-power RF energy harvesting," *IEEE Transaction on Power Electronics*, vol. 23, no. 3, pp. 1494–1501, 2008.

71. R.W. Erickson and D. Maksimovic, *Fundamentals of Power Electronics*, 2nd ed., Springer, New York, pp. 637–663, 2001.

72. J. Twidell and A. Weir, *Renewable Energy Resources*, 2nd ed., Taylor & Francis, London, 2006.

73. The Bor Forest Island Fire Experiment Fire Research Campaign Asia-North (FIRESCAN), *IV. Bor Forest Island Fire Behavior and Atmospheric Emissions*, http://www.fire.uni-freiburg.de/other_rep/research/rus/rus_re_1bor.htm, accessed on July 1, 2010.

74. V. Salas, E. Olias, A. Barrado, and A. Lazaro, "Review of the maximum power point tracking algorithms for stand-alone photovoltaic systems," *Solar Energy Materials and Solar Cells*, vol. 90, no. 11, pp. 1555–1578, 2006.

75. M. Mitchell, "Animated demonstration of Bernoulli's principle," http://home.earthlink.net/mmc1919/venturi.html, accessed on July 1, 2010.

76. S.J. I'Anson, "Radius of curvature," University of Manchester, http://pygarg.ps.umist.ac.uk/ianson/paper_physics/Radius_of_Curvature.html, accessed on July 1, 2010.

77. N.G. Elvin and A.A. Elvin, "A general equivalent circuit model for piezoelectric generators," *Journal of Intelligent Material Systems and Structures*, vol. 20, no. 1, pp. 3–9, 2009.

78. J.G. Smits, S.I. Dalke, and T.K. Cooney, "The constituent equations of piezoelectric bimorphs," *Sensors and Actuators*, vol. A28, no. 1, pp. 41–61, 1991.

79. J.L. Gonzalez, A. Rubio, and F. Moll, "Human powered piezoelectric batteries to supply power of wearables electronic devices," *International Journal of the Society of Materials Engineering for Resources*, vol. 10, no. 1, pp. 34–40, 2002.

80. Q.-M. Wang, X. Du, B. Xu, and L.E. Cross, "Theoretical analysis of the sensor effect of cantilever piezoelectric benders," *Journal of Applied Physics*, vol. 85, no. 3, pp. 1702–1712, 1999.

81. J. Kymissis, C. Kendall, J. Paradiso, and N. Gershenfeld, "Parasitic power harvesting in shoes," *2nd International Symposium on Wearable Computers*, pp. 132–139, 1998.

82. V.C. Gungor and G.P. Hancke, "Industrial wireless sensor networks: challenges, design principles, and technical approaches," *IEEE Transactions on Industrial Electronics*, vol. 56, no. 10, pp. 4258–4265, 2009.

83. C. Alippi, G. Anastasi, M. Di Francesco, and M. Roveri, "Energy management in wireless sensor networks with energy-hungry sensors," *IEEE Transactions on Instrumentation and Measurement Magazine*, vol. 12, no. 2, pp. 16–23, 2009.

84. D.M. Rowe, *Thermoelectrics Handbook: Macro to Nano*, CRC Press, Taylor & Francis, Boca Raton, FL, 2006.

85. J. Carmo, L. Goncalves, and H. Correia, "Thermoelectric micro converter for energy harvesting systems," *IEEE Transaction on Industrial Electronics*, vol. 57, no. 3, pp. 861–867, 2009.

86. W. Glatz, S. Muntwyler, and C. Hierold, "Optimization and fabrication of thick flexible polymer based micro thermoelectric generator," *Sensors and Actuators A: Physical*, vol. 132, pp. 337–345, 2006.

87. N. Femia, G. Petrone, G. Spagnuolo, and M. Vitelli, "Optimization of perturb and observe maximum power point tracking method," *IEEE Transactions on Power Electronics*, vol. 20, no. 4, pp. 963–973, 2005.

88. F. Liu, S. Duan, F. Liu, B. Liu, and Y. Kang, "A variable step size INC MPPT method for PV systems," *IEEE Transactions on Industrial Electronics*, vol. 55, no. 7, pp. 2622–2628, 2008.

89. T. Esram, J.W. Kimball, P.T. Krein, P.L. Chapman, and P. Midya, "Dynamic maximum power point tracking of photovoltaic arrays using ripple correlation control," *IEEE Transactions on Power Electronics*, vol. 21, no. 5, pp. 1282–1291, 2006.

90. R.-Y. Kim, J.-S. Lai, B. York, and A. Koran, "Analysis and design of maximum power point tracking scheme for thermoelectric battery energy storage system," *IEEE Transactions on Industrial Electronics*, vol. 56, no. 9, pp. 3709–3716, 2009.

91. I. Stark, "Invited talk: thermal energy harvesting with thermo life," *International Workshop on Wearable and Implantable Body Sensor Networks (BSN 2006)*, pp. 19–22, 2006.

92. V. Leonov, "Thermal shunts in thermoelectric energy scavengers," *Journal of Electronic Materials*, vol. 38, no. 7, pp. 1483–1490, 2009.

93. S. Dalola, M. Ferrari, V. Ferrari, M. Guizzetti, D. Marioli, and A. Taroni, "Characterization of thermoelectric modules for powering autonomous sensors," *IEEE Transactions on Instrumentation and Measurement*, vol. 58, no. 1, pp. 99–107, 2009.

94. T.S. Paing and R. Zane, "Resistor emulation approach to low-power energy harvesting," *37th IEEE Power Electronics Specialists Conference (PESC)*, pp. 1–7, 2006.

95. J. Sun, D.M. Mitchell, M.F. Greuel, P.T. Krein, and R.M. Bass, "Averaged modeling of PWM converters operating in discontinuous conduction mode," *IEEE Transactions on Power Electronics*, vol. 16, no. 4, pp. 482–492, 2001.

96. V. Vorperian, "Simplified analysis of PWM converters using model of PWM switch. Part II: Discontinuous conduction mode," *IEEE Transactions on Aerospace and Electronic Systems*, vol. 26, pp. 497–505, 1990.

97. P.D. Mitcheson, P. Miao, B.H. Stark, E.M. Yeatman, A.S. Holmes, and T.C. Green, "MEMS electrostatic micropower generator for low-frequency operation," *Sensors and Actuators A: Physical*, vol. 115, nos. 2–3, pp. 523–529, 2004.

98. S. R. Anton and H. A. Sodano, "A review of power harvesting using piezoelectric materials (2003–2006)," *Smart Materials and Structures*, vol. 16, no. 3, pp. R1–R21, 2007.

99. S. Roundy, E.S. Leland, J. Baker, E. Carleton, E. Reilly, E. Lai, B. Otis, J.M. Rabaey, P.K. Wright, and V. Sundararajan, "Improving power output for vibration-based energy scavengers," *IEEE Pervasive Computing*, vol. 4, no. 1, pp. 28–36, 2005.

100. Wikipedia, "Piezoelectricity," http://en.wikipedia.org/wiki/Piezoelectricity, accessed on May 7, 2010.

101. N.S. Shenck, "A demonstration of useful electric energy generation from piezoceramics in a shoe," Master's thesis, Massachusetts Institute of Technology (MIT), Cambridge, MA, 1999.

102. M. Renaud, P. Fiorini, R. Van Schaijk, and C. Van Hoof, "Harvesting energy from the motion of human limbs: The design and analysis of an impact-based piezoelectric generator," *Smart Materials and Structures*, vol. 18, no. 3, (16), 2009.

103. S.P. Beeby, M.J. Tudor, and N.M. White, "Energy harvesting vibration sources for microsystems applications," *Measurement Science and Technology*, vol. 17, no. 12, pp. R175–R195, 2006.

104. M. Umeda, K. Nakamura, and S. Ueha, "Analysis of the transformation of mechanical impact energy to electric energy using piezoelectric vibrator," *Japanese Journal of Applied Physics*, vol. 35, no. 5B, pp. 3267–3273, 1996.

105. C. Keawboonchuay and T.G. Engel, "Maximum power generation in a piezoelectric pulse generator," *IEEE Transactions on Plasma Science*, vol. 31(2), no. 1, pp. 123–128, 2003.

106. M. Renaud, P. Fiorini, and C. Van Hoof, "Optimization of a piezoelectric unimorph for shock and impact energy harvesting," *Smart Materials and Structures*, vol. 16, no. 4, pp. 1125–1135, 2007.

107. J.F. Antaki, G.E. Bertocci, E.C. Green, A. Nadeem, T. Rintoul, R.L. Kormos, and B.P. Griffith, "A gait powered autologous battery charging system for artificial organs," *American Society of Artificial Internal Organs Conference*, pp. M588–M595, 1995.

108. J. Paradiso and M. Feldmeier, "A compact, wireless, self-powered pushbutton controller," *Ubicomp 2001: Ubiquitous Computing, LNCS 2201,* Springer-Verlag, New York, 2001, pp. 299–304.

109. F. Schmidt and M. Heiden, *Wireless Sensors Enabled by Smart Energy—Concepts and Solutions*, EnOcean GmbH, Oberhaching, Germany.

110. A. Rida, L. Yang, and M. Tentzeris, "Chapter 5: State-of-the-art technology for RFID/sensors," *RFID-Enabled Sensor Design and Applications*, Artech House, London, 2010.

111. Piezo Systems, Inc., "Introduction to piezo transducers," http://www.piezo.com/tech2intropiezotrans.html, accessed on May 17, 2010.

112. S.B. Dewan, G.R. Slemon, and A. Straughen, *Power Semiconductor Drives*, Wiley, New York, 1984, Chapters 2 and 5.

113. K.Y. Hoe, "An investigation of self-powered RF wireless sensors," Bachelor's thesis, National University of Singapore, 2006.

114. H.W. Kim, A. Batra, S. Priya, K. Uchino, D. Markley, R.E. Newnham, and H.F. Hofmann, "Energy harvesting using a piezoelectric cymbal transducer in dynamic environment," *Japanese Journal of Applied Physics, Part 1 (Regular Papers, Short Notes and Review Papers)*, vol. 43, no. 9A, pp. 6178–6183, 2004.

115. B.R. Face, "Self-powered, electronic keyed, multifunction switching system," Face International, Patent U.S. 7161276, 2007.

116. D. Dausch and S. Wise, *Compositional Effects on Electromechanical Degradation of RAINBOW Actuators, NASA*, Hampton, VA, 1998.

117. R.G. Bryant, "LaRC^{TM}-SI: a soluble aromatic polyimide," *High Performance Polymers*, vol. 8, pp. 607–615, 1996.

118. K. Mossi, C. Green, Z. Ounaies, and E. Hughes, "Harvesting energy using a thin unimorph prestressed bender: Geometrical effects," *Journal of Intelligent Material Systems and Structures*, vol. 16, no. 3, pp. 249–261, 2005.

119. K. Mossi, Z. Ounaies, and S. Oakley, "Optimizing energy harvesting of a composite unimorph pre-stressed bender," *16th Technical Conference of the American Society for Composites*, 2001.

120. A.D. Danak, H.S. Yoon, and G.N. Washington, "Optimization of electrical output in response to mechanical input in piezoceramic laminated shells," *ASME International Congress and Exposition*, pp. 309–315, 2003.

121. J.T. Cher, "Design of a piezoelectric energy harvesting wireless electronic switch," Bachelor's thesis, National University of Singapore, 2008.

122. M.K. Stojcev, M.R. Kosanovic, and L.R. Golubovic, "Power management and energy harvesting techniques for wireless sensor nodes," *9th International Conference on Telecommunications in Modern Satellite, Cable, and Broadcasting Services*, pp. 65–72, 2009.

123. S. Taylor, N. Miller, W. Sifuentes, E. Moro, G. Park, C. Farrar, E. Flynn, D. Mascarenas, and M. Todd, "Energy harvesting and wireless energy transmission for embedded sensor nodes," *Proceedings of the SPIE—The International Society for Optical Engineering*, vol. 7288, p. 728810, 2009.

124. R. Torah, P. Glynne-Jones, M. Tudor, T. O'Donnell, S. Roy, and S. Beeby, "Self-powered autonomous wireless sensor node using vibration energy harvesting," *Measurement Science and Technology*, vol. 19, no. 12, p. 125202, 2008.

125. Y.T. He, Y.Q. Li, L.H. Liu, and L. Wang, "Solar micro-power system for self-powered wireless sensor nodes," *Proceedings of the SPIE—The International Society for Optical Engineering*, vol. 7133, p. 71333Z (8 pp.), 2008.

126. Y. Tadesse, S. Zhang, and S. Priya, "Multimodal energy harvesting system: piezoelectric and electromagnetic," *Journal of Intelligent Material Systems and Structures*, vol. 20, no. 5, pp. 625–632, 2009.

127. A. Khaligh, P. Zeng, and C. Zheng, "Kinetic energy harvesting using piezoelectric and electromagnetic technologies—State of the art," *IEEE Transaction on Industrial Electronics*, vol. 57, no. 3, pp. 850–860, 2010.

128. N.J. Guilar, T.J. Kleeburg, A. Chen, D.R. Yankelevich, and R. Amirtharajah, "Integrated solar energy harvesting and storage," *IEEE Transactions on Very Large Scale Integration (VLSI) Systems*, vol. 17, no. 5, pp. 627–637, 2009.

129. H. Lhermet, C. Condemine, M. Plissonnier, R. Salot, P. Audebert, and M. Rosset, "Efficient power management circuit: from thermal energy harvesting to above-IC microbattery energy storage," *IEEE Journal of Solid-State Circuits*, vol. 43, no. 1, pp. 246–255, 2008.

130. A.N. Celik and N. Acikgoz, "Modelling and experimental verification of the operating current of mono-crystalline photovoltaic modules using four- and five-parameter models," *Applied Energy*, vol. 84, no. 1, pp. 1–15, 2007.

131. M.G. Villalva, J.R. Gazoli, and E.R. Filho, "Comprehensive approach to modeling and simulation of photovoltaic arrays," *IEEE Transaction on Power Electronics*, vol. 24, no. 5, pp. 1198–1208, 2009.

132. D. Sera, R. Teodorescu, and P. Rodriguez, "PV panel model based on datasheet values," *IEEE International Symposium on Industrial Electronics (ISIE)*, pp. 2392–2396, 2007.

133. T. Esram and P.L. Chapman, "Comparison of photovoltaic array maximum power point tracking techniques," *IEEE Transactions on Energy Conversion*, vol. 22, no. 2, pp. 439–449, 2007.

134. R. Faranda and S. Leva, "Energy comparison of MPPT techniques for PV Systems," *WSEAS Transactions on Power Systems*, vol. 3, no. 3, pp. 446–455, 2008.

135. K.H. Hussein, I. Muta, T. Hoshino, and M. Osakada, "Maximum photovoltaic power tracking: An algorithm for rapidly changing atmospheric conditions," *IEE Proceedings—Generation, Transmission and Distribution*, vol. 142, no. 1, pp. 59–64, 1995.

136. D. Brunelli, L. Benini, C. Moser, and L. Thiele, "An efficient solar energy harvester for wireless sensor nodes," *Design, Automation and Test in Europe*, pp. 104–109, 2008.

137. Y.Q. Li, H.Y. Yu, B. Su, and Y.H. Shang, "Hybrid micropower source for wireless sensor network," *IEEE Sensors Journal*, vol. 8, no. 6, pp. 678–681, 2008.

138. D. Dudek, C. Haas, A. Kuntz, M. Zitterbart, D. Krger, P. Rothenpieler, D. Pfisterer, and S. Fischer, "A wireless sensor network for border surveillance," *Proceedings of the 7th ACM Conference on Embedded Networked Sensor Systems*, pp. 303–304, 2009.

139. Solar4Power, "Global solar power map #2: North America from Canada to Texas," http://www.solar4power.com/map2-global-solar-power.html, accessed on June 15, 2010.

140. *Canadian Wind Energy Atlas*, http://www.windatlas.ca/en/index.php, accessed on June 15, 2010.

141. Osram, "Special lamps: Osram lamp technology," www.friarsmarketing.com/Resources/SPECIAL%20LAMPS.pdf, accessed on June 15, 2010.

142. Q.Y. Liu, "Hybrid energy harvesting from wind and solar energy sources to power wireless sensor nodes," Bachelor's thesis, National University of Singapore, 2010.

143. A. Nasiri, S.A. Zabalawi, and G. Mandic, "Indoor power harvesting using photovoltaic cells for low-power applications," *IEEE Transactions on Industrial Electronics*, vol. 56, no. 11, pp. 4502–4509, 2009.

144. A. Hande, T. Polk, W. Walker, and D. Bhatia, "Indoor solar energy harvesting for sensor network router nodes," *Microprocessors and Microsystems*, vol. 31, no. 6, pp. 420–432, 2007.

145. J.F. Randall, *"Designing Indoor Solar Products: Photovoltaic Technologies for AES,"* Wiley, Hoboken, NJ, 2005.

146. A. Wickenheiser and E. Garcia, "Combined power harvesting from AC and DC sources," *Proceedings of the SPIE*, vol. 7288, p. 728816-1-9, 2009.

147. A.S. Marincic, "Nikola Tesla and the wireless transmission of energy," *IEEE Transactions on Power Apparatus and Systems*, vol. PAS-101, no. 10, pp. 4064–4068, 1982.

148. S. Ahson and M. Ilyas, *"RFID Handbook: Applications, Technology, Security, and Privacy,"* CRC Press, Boca Raton, FL, 2008.

149. A. Sample, D. Yeager, P. Powledge, A. Mamishev, and J. Smith, "Design of an rfid-based battery-free programmable sensing platform," *IEEE Transactions on Instrumentation and Measurement*, vol. 57, no. 11, pp. 2608–2615, 2008.

150. W. Brown, "The history of power transmission by radio waves," *IEEE Transactions on Microwave Theory and Techniques*, vol. 32, no. 9, pp. 1230–1242, 1984.

151. J. McSpadden and J. Mankins, "Space solar power programs and microwave wireless power transmission technology," *IEEE Microwave Magazine*, vol. 3, no. 4, pp. 46–57, 2002.

152. A. Sample and J. Smith, "Experimental results with two wireless power transfer systems," *IEEE Radio and Wireless Symposium*, pp. 16–18, 2009.

153. Z.N. Low, R.A. Chinga, R. Tseng, and J.S. Lin, "Design and test of a high-power high-efficiency loosely coupled planar wireless power transfer system," *IEEE Transactions on Industrial Electronics*, vol. 56, no. 5, pp. 1801–1812, 2009.

154. P. Sample, T. Meyer, and R. Smith, "Analysis, experimental results, and range adaptation of magnetically coupled resonators for wireless power transfer," *IEEE Transactions on Industrial Electronics*, pp. 1–11, 2010.

155. PowerMat Inc., http://www.powermat.com, accessed on May 20, 2010.

156. C. Zhu, K. Liu, C. Yu, R. Ma, and H. Cheng, "Simulation and experimental analysis on wireless energy transfer based on magnetic resonances," *IEEE Vehicle Power and Propulsion Conference*, pp. 1–4, 2008.

157. Z.N. Low, R. Chinga, R. Tseng, and J. Lin, "Design and test of a high-power high-efficiency loosely coupled planar wireless power transfer system," *IEEE Transactions on Industrial Electronics*, vol. 56, no. 5, pp. 1801–1812, 2009.

158. J. Casanova, Z.N. Low, and J. Lin, "A loosely coupled planar wireless power system for multiple receivers," *IEEE Transactions on Industrial Electronics*, vol. 56, no. 8, pp. 3060–3068, 2009.

159. B. Jiang, J.R. Smith, M. Philipose, S. Roy, K. Sundara-Rajan, and A.V. Mamishev, "Energy scavenging for inductively coupled passive RFID systems," *IEEE Transactions on Instrumentation and Measurement*, vol. 56, no. 1, pp. 118–125, 2007.

160. Y. Yang, D. Divan, R.G. Harley, and T.G. Habetler, "Power line sensornet—A new concept for power grid monitoring," *IEEE Power Engineering Society General Meeting*, pp. 1–8, 2006.

161. G.X. Wang, W.T. Liu, M. Sivaprakasam, and G.A. Kendir, "Design and analysis of an adaptive transcutaneous power telemetry for biomedical implants," *IEEE Transactions on Circuits and Systems I: Regular Papers*, vol. 52, no. 10, pp. 2109–2117, 2005.

162. J.T. Boys, G.A.J. Elliott, and G.A. Covic, "An appropriate magnetic coupling coefficient for the design and comparison of ICPT pickups," *IEEE Transactions on Power Electronics*, vol. 22, no. 1, pp. 333–335, 2007.

163. A. Kurs, A. Karalis, R. Moffatt, J.D. Joannopoulos, P. Fisher, and M. Soljacic, "Wireless power transfer via strongly coupled magnetic resonances," *Science Magazine*, vol. 317, no. 5834, pp. 83–86, 2007.

164. D. Penly, "Induction," http://facstaff.gpc.edu/dpenly/1112/Induct.pdf, accessed on May 19, 2010.

165. S.M. Kanbur, "Induction and inductance," http://www.oswego.edu/~kanbur/phy313/Fard.pdf, accessed on May 19, 2010.

166. Y.K. Tan and S.K. Panda, "A novel method of harvesting wind energy through piezoelectric vibration for low-power autonomous sensors," *nanoPower Forum (nPF'07)*, 2007.

167. S.C. Xie, "Inductive energy transfer system," Bachelor's thesis, National University of Singapore, 2008.

168. "Wireless power supply," http://www.wirelesspowersupply.net/, accessed on May 26, 2010.

169. D. Murphy, "Case-mate's hug wireless iPhone charging solution shipping now for $100," http://www.engadget.com/2010/03/09/case-mates-hug-wireless-iphone-charging-solution-shipping-now-f/, accessed on May 26, 2010.

170. "Sony develops effective wireless power transmission up to 60W," http://www.mydigitallife.info/2009/10/03/sony-develops-effective-wireless-power-transmission-up-to-60w/comment-page-1/, accessed on May 26, 2010.

171. L. Jorgensen and A. Culberson, "Wireless power transmission using magnetic resonance," http://www.cornellcollege.edu/physics/courses/phy312/Student-Projects/Magnetic-Resonance/Magnetic-Resonance.html, accessed on May 26, 2010.

172. P. Somasundaram, "Analysis and optimization of strongly coupled magnetic resonance for wireless power transfer applications," Bachelor's thesis, National University of Singapore, 2010.

173. W.J. Chow, "Wireless transmission of power with magnetic resonance," Bachelor's thesis, National University of Singapore, 2008.

174. A. Karalis, J. Joannopoulos, and M. Soljacic, "Efficient wireless non-radiative mid-range energy transfer," *ScienceDirect-Annals of Physics*, vol. 323, no. 1, pp. 34–48, 2008.

175. B. Cannon, J. Hoburg, D. Stancil, and S. Goldstein, "Magnetic resonant coupling as a potential means for wireless power transfer to multiple small receivers," *IEEE Transactions on Power Electronics*, vol. 24, no. 7, pp. 1819–1825, 2009.

Index